Farm Hall and the German Atomic Project of World War II

Farm Hall 1945 (NARA, Farm Hall Reports, courtesy of AIP ESVA)

David C. Cassidy

Farm Hall and the German Atomic Project of World War II

A Dramatic History

 Springer

David C. Cassidy
Hempstead, NY
USA

ISBN 978-3-319-86654-3   ISBN 978-3-319-59578-8   (eBook)
DOI 10.1007/978-3-319-59578-8

Acknowledgments

I am very grateful to H. Frederick Dylla, Director Emeritus of the American Institute of Physics (AIP); to Gregory A. Good, Director of the AIP Center for History of Physics; and to R. Joseph Anderson, Director Emeritus and Melanie Mueller, Director, of the AIP Niels Bohr Library and Archives, for their generous support and encouragement in the preparation of this book.

I wish to express my sincere appreciation to Prof. Brian Schwartz for his suggestion that I attempt to write the play *Farm Hall*. The play was developed in collaboration with Break A Leg Productions, New York and its artistic director Teri Black, to whom I am very grateful for their patience and support. I am also very grateful to director and dramaturge Jean Dobie Giebel, and to Brian Schwartz, Gerald Holton and Ruth Lewin Sime for their very helpful comments and suggestions. I benefitted greatly from private and staged readings presented by Break A Productions at the CUNY Graduate Center, New York; the Baltimore Theatre Group (Norman Seltzer, director), sponsored by the Forum on the History of Physics, during a meeting of the American Physical Society; and Dream Catchers, New York. I thank Prof. Peter Pesic and the students of St. John's College, Santa Fe, New Mexico who presented a public reading of the play sponsored by the AIP that ultimately led to this book. I am sincerely grateful to director Gerald vanHeerden and to the actors and staff of Break A Leg Productions who brought this play to the stage; to Spencer R. Weart and Benjamin Bederson for their post-performance commentary on the play; and to John Chatterton, director of the Midtown International Theatre Festival, New York. I very much appreciate the excellent advice and hard work of all of the actors and directors involved over the years in the development of the play.

I would also like to express my appreciation to Melanie Mueller, Director, and Audrey Lengel, Photograph Librarian of the AIP Niels Bohr Library and Archives, for assistance in obtaining the photographs for this book. I am grateful to Michaele Thurgood Haynes, Terry Thurgood, and the other rights holders for their permission to publish these materials.

Contents

Abbreviations

AIP   American Institute of Physics, College Park, Maryland
ESVA   Emilio Segrè Visual Archives, at the AIP
NARA   US National Archives and Records Administration, College Park, Maryland
NBLA   Niels Bohr Library and Archives, at the AIP

Introduction

As the Allied armies fought their way into Germany following the D-Day invasion in 1944, two teams of German nuclear scientists hastened to complete their work. The first, led by the world-renowned quantum physicist Werner Heisenberg, had evacuated war-torn Berlin for the quiet hills of southwestern Germany. There he and his colleagues made their last attempt to achieve a self-sustaining nuclear reactor. A second, smaller reactor project, headed by applied physicist Kurt Diebner and supported in part by the German Army and the SS, had relocated to the Thüringen Forest farther north.

Both German efforts labored under the conviction that they were far ahead of those they called the "Anglo-Americans" in nuclear matters. If they could not build a reactor, let alone a nuclear explosive, they thought, then neither could the Anglo-Americans. Little did they know that the Manhattan Project, one of the largest technological research projects ever launched, was at that time feverishly preparing the atomic bombs to be dropped on Germany (until Germany surrendered in May) and on Japan in August 1945.

As they raced to build the atomic bombs, the Anglo-Americans likewise believed the Germans to be far ahead. Not knowing how far the Germans had progressed, the commander of the Manhattan Project, Major General Leslie R. Groves, dispatched into Europe a secret American science intelligence unit, the Alsos Mission, with Dutch-American physicist Samuel A. Goudsmit as its scientific head. (*álsos* is Greek for "grove.") Later joined by British agents and officers, its mission was to follow close behind the advancing Allied forces—first in Italy, then in Germany—to locate and disable German nuclear fission research, and to capture and detain the scientists involved, along with their papers and equipment. Before either German project could succeed in its quest for a reactor, the first step toward a bomb, Goudsmit and his mission, backed by a combat engineer battalion, swept in and captured the scientists during the last days of the war in Europe. They brought the captives, together with their documents, to Alsos headquarters in Heidelberg for interrogation, then held them for two months in a series of prisoner-of-war camps in France and Belgium. Their equipment was dismantled and sent to an American base near Paris.

The Anglo-Americans selected ten of their captive nuclear scientists for long-term detention in British custody, keeping them off the Continent and away from the Russians, the French, and any curious reporters. Among the captives were Nobel Prize winners Werner Heisenberg, Max von Laue, and Otto Hahn (who received his prize while in custody), as well as Kurt Diebner, Walther Gerlach, and Carl Friedrich von Weizsäcker. On July 3, 1945 the British brought the selected scientists to an English country manor, Farm Hall, a MI-6 safe house near Cambridge. The house and nearby airport had been used during the war in support of clandestine operations behind enemy lines. The scientists remained at Farm Hall until the British returned them to Germany exactly six months later, on January 3, 1946, and released them within the British occupation zone in northwestern Germany.

Unknown to the scientists, British agents recorded their conversations while at Farm Hall using hidden microphones installed throughout the house and grounds. The bilingual agents transcribed and translated selected portions of the conversations for their intelligence interest. They worked under the direction of British Army Major T.H. Rittner and, when Rittner became ill, under his deputy, Captain P.L.C. Brodie. Rittner or Brodie periodically transmitted reports of the scientists' conversations, including extensive excerpts from the English translations and some original German passages, to General Groves and to British authorities. The top-secret reports, amounting to 153 pages, remained classified until 1992, when British and American officials jointly released them to the public. Unfortunately, the original recordings were destroyed when the agents reused the shellacked metal discs on which the recordings were made. No additional transcriptions of the recorded conversations have been found.

The Farm Hall captivity occurred during a profound turning point in history, from conventional warfare to the nuclear age and from brutal dictatorship and the Holocaust to the nuclear arms race and the Cold War clash between East and West. It is a well-researched—and widely written about—period of contemporary history and history of science, and yet, regarding Germany's nuclear program, it remains one of the most controversial, periodically engendering heated debates concerning the German scientists' relative lack of success and their motives, aims and rationales for working on nuclear fission under the Hitler regime at war.

The scientists' six-month captivity at Farm Hall also encompassed the challenging periods before, during, and after the atomic bombings of Japan. With most of Europe in ruins, the economy and infrastructure in collapse, entire populations on the verge of starvation, the scientists experienced these events as if on another planet, ensconced in the isolated comfort and care of Farm Hall. The situation is even more extraordinary in that we have the actual verbatim conversations of the scientific participants on the German side as these events are unfolding. We can follow in the transcripts of their apparently unguarded conversations their reactions to the German defeat; their anguished response to the news of the atomic bombs; their dashed illusions of progress beyond Allied nuclear efforts; their difficult adjustment to the early nuclear age; their concerns for their reputations; and their concerns for their families left behind in war-torn Germany. How had the

Americans and British done it? Why had they themselves made such little progress in comparison? Why are they being held? What will happen to them? How can they explain their war-time work and their meager progress to their countrymen, to their former enemies, and to themselves? While a historical account can offer well-reasoned, source-based answers to these questions, a theatrical play allows us to explore behind the events, to experience with the characters their encounter with these momentous events, with the issues they raised, and with each other. It is at the intersection of the historical events with the human qualities, motives, and emotions of real people that both the drama and the history come alive, enabling us to encounter and appreciate both from new and richer perspectives.

Since the end of World War II, a great many works about the German atomic project and its leading figures have appeared. Among the first was Goudsmit's 1947 memoir of his mission and his assessment of the scientists, simply titled *Alsos*. That book and nearly every subsequent major work or release over the years has unleashed another round of fervent controversy—from Robert Jungk's *Brighter than a Thousand Suns* (1956/58), to General Groves's memoir *Now It Can Be Told* (1962), to the release and publication of the Farm Hall transcripts (1992), to Michael Frayn's popular award-winning play *Copenhagen* (1998). Intense feelings and convictions have obviously become associated with many of these issues, while the dramatic elements of mounting conflict and crisis have pervaded the Farm Hall experience from beginning to end.

The publication of the Farm Hall transcripts in 1992, with their many pages of verbatim dialogue, encouraged several writers to attempt to turn them into a dramatic play. Others, such as myself, became entangled in the historical controversies revived by the transcripts and by Michael Frayn's subsequent play *Copenhagen*. It was a play that explored the related topics of Heisenberg's motives for working on the German project and for visiting Niels Bohr in Copenhagen in 1941 at a crucial moment for the German effort. Although the story of the capture, internment, and release of the Farm Hall scientists is certainly full of historical and dramatic potential, the task of creating a Farm Hall play proved more difficult than it appeared, especially for those new to this genre. Fifteen years later, in 2007, a colleague, Brian Schwartz, approached me during a meeting of the American Physical Society with the observation that no one had yet succeeded in creating a Farm Hall play. He suggested that I give it a try. I knew the history well, he argued, and, with the dialogue at hand, I thought it would not be difficult. How wrong I was! Although I had earlier dabbled in creative writing, I had no idea how to write a play. Given the highly controversial nature of the subject, combined with the required attention to the necessary dramatic elements, it was no easy task to write a successful first play on this subject. With the help of the dramaturge and director Prof. Jean Dobie Giebel at Hofstra University and the encouragement and advice of the New York theatre company Break A Leg Productions and its artistic director Teri Black, as well as the advice of colleagues, audiences, and actors, *Farm Hall* gradually came to life. Following several staged readings, Break A Leg Productions presented the play's world premiere production during the Midtown International Theatre Festival held in New York in July 2014. The play received two festival award nominations for Best Supporting Actor, which went to one of the play's nominees.

In 2013 the Nora Theatre Company in Cambridge, Massachusetts presented *Operation Epsilon* by Alan Brody, a play on the same subject which took a somewhat different approach.

In 2015 the AIP Niels Bohr Library and Archives received from the family of photographer Malcolm Thurgood, a former US Army sergeant assigned to the Alsos Mission, a collection of his photographs taken during his assignment with the mission. These now constitute a valuable supplement to the Goudsmit Papers, held by the AIP, and a welcome visual aid to this and other works of all sorts on these events

This book grew out of an invitation to the author to present the 2014 Lyne Starling Trimble Lecture, sponsored by the AIP Center for History of Physics, on a topic of my choice in the history of physics during a public meeting of the AIP in Santa Fe, New Mexico. Then in the throes of *Farm Hall*, I made the tentative suggestion that, instead of a lecture, I could present the play as a staged reading with local actors, followed by a discussion with the author and actors. To my pleasant surprise the AIP bravely accepted. Prof. Peter Pesic at St. John's College in Santa Fe kindly arranged for students from the college to direct and present a reading of the play, and Brian Schwartz worked tirelessly as producer to arrange publicity and other matters. The students presented a public reading of the play at the New Mexico School for the Deaf on May 7, 2014. Its success, owing above all to the truly outstanding student company, encouraged Frederick Dylla, then AIP director, to propose a publication that would bring together, in what would be a nearly unique work, the three diverse genres of history, drama, and photography: the one-act play *Farm Hall*; a historical survey of the events in the play and their background; and a selection of photographs from the newly acquired Thurgood collection, along with other related photos and archival materials. Together, these three genres illustrate how each can provide very different, yet complementary, perspectives on these events, bringing them into richer and deeper focus than individually possible. This book is also intended to provide readers with an introduction to the wealth of information available in a wide range of formats and genres on the German nuclear project and on the atomic scientists at the dawn of the atomic age.

Farm Hall

It is the end of World War II in Europe. German nuclear scientists are held captive in the English country manor Farm Hall as atomic bombs are on their way to Japan. Based on a true story.

Farm Hall was developed in collaboration with Break A Leg Productions, New York. It was originally presented in July 2014 at the June Havoc Theater in New York, as part of the Midtown International Theatre Festival, and directed by Gerald vanHeerden

Cast

Samuel A. Goudsmit	Kevin Hauver
Werner Heisenberg	Keith Herron
Carl Friedrich von Weizsäcker	James Reade Venable
Major T.H. Rittner	Terrence Keene
Walter Gerlach/BBC Announcer	Scott Glascock
Kurt Diebner	Terrence Montgomery
Otto Hahn	Miller Lide
Voice of Doom	Bruce Barton
Elisabeth Heisenberg	Susan Richard

In memory of Scott Glascock, a great actor and a wonderful person.

Miller Lide won Best Supporting Actor during the Festival for his role as Otto Hahn. Terrence Montgomery was also nominated for that award for his role as Kurt Diebner.

© Springer International Publishing AG 2017
D.C. Cassidy, *Farm Hall and the German Atomic Project of World War II*,
DOI 10.1007/978-3-319-59578-8_1

Characters in Order of Appearance

The approximate pronunciations of names are indicated.

SAMUEL A. GOUDSMIT (pronounced Gaud-smit, his preferred pronunciation). Dutch-American physicist who was a friend and colleague of Heisenberg during the glory days of quantum mechanics in the 1920s. He is now scientific head of the Alsos Mission that captured the German nuclear scientists and halted their fission research. (See photo, page 62)

Werner Heisenberg at Farm Hall (NARA, Farm Hall Reports, courtesy of AIP ESVA)

WERNER HEISENBERG (Verner Hi-zin-berg). Charismatic, youthful Nobel Prize-winning quantum physicist, scientific head of the main German uranium project.

C. F. von Weizsäcker at Farm Hall (NARA, Farm Hall Reports, courtesy of AIP ESVA)

CARL FRIEDRICH VON WEIZSÄCKER (Veits-say-ker). A former student and now close colleague of Heisenberg. His father was an official in the Reich's Foreign Office and was arrested and tried at Nuremberg.

MAJOR T.H. RITTNER. The British intelligence officer in charge at Farm Hall, author of the Farm Hall intelligence reports on the scientists' conversations. (See photo, page 87}

Walther Gerlach at Farm Hall (NARA, Farm Hall Reports, courtesy of AIP ESVA)

WALTHER GERLACH (Valter [silent h] Gaer-lock). A distinguished experimental physicist, best known for the Stern-Gerlach effect in quantum mechanics. As head of the physics section of the Reich Research Council, he was the last administrator of the German nuclear effort.

Kurt Diebner at Farm Hall (NARA, Farm Hall Reports, courtesy of AIP ESVA)

KURT DIEBNER (Deeb-ner). An inventive applied nuclear physicist who helped initiate the German fission project in 1939 through the Army Ordnance (Weapons) Bureau. Later he worked with the support of Gerlach's organization on his own reactor project that rivaled Heisenberg's. Unlike the other scientists, he is not an elite professor.

Otto Hahn at Farm Hall (NARA, Farm Hall Reports, courtesy of AIP ESVA)

OTTO HAHN (Han). Distinguished nuclear chemist, co-discoverer of nuclear fission who says he maintained his distance from the nuclear effort.

ELISABETH HEISENBERG. Heisenberg's wife and the mother of their six children. She is struggling heroically for her family's survival during the post-war chaos in Germany. (See photo, page 40)

VOICE OF DOOM

BBC RADIO ANNOUNCER (British accent)

The Play

<u>SCENE 1</u>

An office. Goudsmit is present. Heisenberg enters.

GOUDSMIT:	Greetings, Heisenberg. So we meet again. (*They shake hands.*)
HEISENBERG:	Goudsmit—my favorite Dutch colleague.
GOUDSMIT:	Dutch-American colleague.
HEISENBERG:	Yes, of course.
GOUDSMIT:	Well, it's been a while.
HEISENBERG:	And much has happened.
GOUDSMIT:	Do you remember our last meeting?
HEISENBERG:	Summer 1939. Just before the outbreak of war in Europe.
GOUDSMIT:	You stayed with us during your lectures at Michigan.
HEISENBERG:	We talked for hours into the night about Copenhagen, and quantum mechanics, and Niels Bohr.... Those were the days.
GOUDSMIT:	And do you remember how everyone there tried to convince you to stay and work with us on nuclear fission?
HEISENBERG:	Yes.
GOUDSMIT:	Don't you want to come to America now and work with us?
HEISENBERG:	No. I don't want to leave. Germany needs me.
GOUDSMIT:	That's just what you said back then!
HEISENBERG:	It's even more true now. Germany's in ruins. The bombing destroyed everything. The economy's in collapse. And my institute needs me. My family needs me. They're all in a small cabin in the mountains with barely any food.
GOUDSMIT:	And <u>my</u> family needed you!
HEISENBERG:	You mean—
GOUDSMIT:	My parents.
HEISENBERG:	How are they?
GOUDSMIT:	You don't know?
HEISENBERG:	Dirk Coster wrote from The Hague that your parents had been arrested.
GOUDSMIT:	(*angry*) And all you did was write some vague letter to Coster about how hospitable I was to you in Ann Arbor, how concerned you would be if my parents were subjected to any "difficulties."
HEISENBERG:	I did what I could.
GOUDSMIT:	(*shouting*) But you took a month to respond with such a letter. My Jewish parents were already sent to Auschwitz!
HEISENBERG:	What! I didn't know it was so soon.
GOUDSMIT:	But you did know what that meant.
HEISENBERG:	We heard many rumors during the war.
GOUDSMIT:	Now you're being vague again!

HEISENBERG: It's difficult to forget the lessons one learns for survival.

GOUDSMIT: So why didn't you respond sooner?

HEISENBERG: I did respond—as soon as I could. I thought your parents were still in The Hague. So I thought Coster could use my letter to obtain their release. Obviously it was too late.

GOUDSMIT: Damned right it was too late! My father and my blind mother were murdered in Auschwitz a month later. It was my father's 70th birthday!

HEISENBERG: Oh, no!

GOUDSMIT: (*shouting*) You were the leading scientist on Germany's nuclear fission project! You had authority! You had connections in high places!

HEISENBERG: They weren't enough. Nobody listened to us. We were just physicists trying to survive. What could we do against the Third Reich at war?

GOUDSMIT: But you were damned good physicists with a potentially big contribution to the war effort! You could have used that. You could have done something. You could have gone to Weizsäcker's father in the Foreign Ministry—anything but writing some vague bureaucratic letter!

HEISENBERG: I—I guess you're right.... But I had to be vague. I was under surveillance. They opened my letters.

GOUDSMIT: We were friends! You met my parents!

HEISENBERG: Yes...You're right. I—I'm truly sorry.

GOUDSMIT: Excuse me.

Goudsmit exits. Weizsäcker enters.

HEISENBERG: Weizsäcker! Thank goodness you're here.

WEIZSÄCKER: Are you alright?

HEISENBERG: Goudsmit's parents died in Auschwitz.

WEIZSÄCKER: The letter to Coster didn't help?

HEISENBERG: He blames me for not acting sooner and then sending only a vague letter.

WEIZSÄCKER: You can't be blamed for what others have done.

HEISENBERG: Maybe I could have done more. Maybe I could have gone to Göring as head of the Aviation Academy. We could have visited your father in the Foreign Office.

WEIZSÄCKER: But you know how difficult things were. The SS didn't allow anyone meddling in their work.

HEISENBERG: Still, I think of what more we could have done.

WEIZSÄCKER: Heisenberg, we have to look to the future now. We can't keep dwelling on the past.

HEISENBERG: I suppose you're right.

WEIZSÄCKER: Germany needs us.

HEISENBERG:	At least we have one advantage over the Allies.
WEIZSÄCKER:	What's that?
HEISENBERG:	Our uranium research has surely put Germany far ahead of anything the Allies could have achieved.
WEIZSÄCKER:	The Anglo-Americans will have to come to us if they want to build a reactor anytime soon.
HEISENBERG:	Our success should also help us shape the future of German science.
WEIZSÄCKER:	And help us leave the past far behind.
HEISENBERG:	I just hope others will do the same.
WEIZSÄCKER:	(derisive) Yes—Diebner.
HEISENBERG:	He never did approve of my leading role.
WEIZSÄCKER:	And—he was a Nazi party member.
HEISENBERG:	Well, let's hope we all can put the past far behind us.

SCENE 2

Goudsmit and Rittner are present.

RITTNER:	(to audience) Major T.H. Rittner, British Army Intelligence, report. On May 3, 1945 Dr. Goudsmit and his science intelligence mission captured Prof. Werner Heisenberg, the last of the targeted German nuclear scientists. They completely dismantled his uranium project in southern Germany. The scientists were brought to Heidelberg, Germany for interrogation. From there they were transported under close guard to a prisoner of war camp near Versailles and then to one near Brussels, Belgium. Pursuant to orders, I met with Dr. Goudsmit in Brussels on June 14.

Rittner joins Goudsmit.

GOUDSMIT:	Major Rittner, glad to meet you. (*They shake hands.*)
RITTNER:	Same here, Dr. Goudsmit.
GOUDSMIT:	Major, as you know, during the past 6 weeks the Americans have been holding 23 German nuclear scientists in our prisoner-of-war camps.
RITTNER:	Yes.
GOUDSMIT:	We asked your superiors to send their best officer to meet with me here to discuss our plans.
RITTNER:	Glad to be of service.
GOUDSMIT:	Glad to have you. Do you speak German?
RITTNER:	Jawohl [*Ya vohl*].
GOUDSMIT:	Excellent…Well, our officials have decided to remove the most important of the German nuclear scientists from the Continent.
RITTNER:	To Britain?

GOUDSMIT: (*nods*) We've already selected the top 5 for removal. Four of them are professors.

RITTNER: German professors! They <u>are</u> important.

GOUDSMIT: We'll transport them by military aircraft from Brussels to the airbase near Cambridge, then bring them to Farm Hall, an old country manor now serving as an MI-6 safe house.

RITTNER: I know it well.

GOUDSMIT: Major, you will be in charge of the scientists during the journey and while they are detained at Farm Hall. While at Farm Hall the scientists are to communicate with absolutely no one.

RITTNER: That won't be easy. How long will this be?

GOUDSMIT: I have no idea. And one more thing, Major. While they are at Farm Hall you and your staff are to provide periodic reports to the commander of the Manhattan Project on the scientists' loyalties and behavior—and on their nuclear research work.

RITTNER: Dr. Goudsmit, are you sure this is wise?

GOUDSMIT: What do you mean?

RITTNER: I don't know a thing about nuclear physics. What can I possibly report?

GOUDSMIT: Everything they have to say about their work.

RITTNER: But they're nuclear scientists!

GOUDSMIT: And one of them has a Nobel Prize in physics.

RITTNER: Now I'm really over my head. Don't you think someone else—

GOUDSMIT: Don't worry, Major, you'll do fine. (*Pats him on the back*)

RITTNER: But Farm Hall? It's not a secure location. What's to prevent them from escaping?

GOUDSMIT: I understand you were born in Germany.

RITTNER: Yes.

GOUDSMIT: How long did you live there?

RITTNER: I was 12 when I left.

GOUDSMIT: Do you happen to remember the high value German academics place on keeping their word at all costs? It's a matter of personal honor.

RITTNER: This is going to be interesting.

SCENE 3

Comfortable living room. Refreshments on a table. Gerlach, Weizsäcker, Diebner, and Hahn enter looking around.

GERLACH: This is nice.

WEIZSÄCKER: My goodness, what comfort. Heisenberg will be pleased.

GERLACH: Too bad we couldn't all fit in the car.

DIEBNER:	The next trip from the airbase will be a sight—the Nobel Prize winner squeezed between the kitchen staff!
HAHN:	Look over here. (*He finds a paper on a table.*) They've assigned us bedrooms. Let's see. Gerlach and Weizsäcker, you two are in the first room. I'm alone in the second room. And—do you believe it!—Diebner, you're with Heisenberg in the room at the end of the hall.
DIEBNER:	I wonder how he'll react to that.
HAHN:	It says here the other rooms are for the intelligence staff.
DIEBNER:	I didn't know intelligence needed a staff.
HAHN:	Apparently theirs does.
GERLACH:	This is definitely a quantum leap up from that Belgian POW camp. A month there seemed like an eternity.
WEIZSÄCKER:	(*bitter*) They called it Dustbin! They even put us in with common soldiers.
GERLACH:	Farm Hall sounds much better.
HAHN:	The English are treating us like royalty. We have comfortable quarters, German orderlies, all of the English newspapers and books we could want.
GERLACH:	Even some refreshments. (*He reaches to take some.*)
HAHN:	Careful, Gerlach! They may be poisoned.
GERLACH:	(*He quickly pulls his hand back.*) You think so?
HAHN:	I'm only kidding.
WEIZSÄCKER:	But maybe they did bring us here just so they can kill us off quietly.
HAHN:	All this way to feed us poisoned food in the lap of luxury?
DIEBNER:	Hahn, you're always the most sensible.
HAHN:	I'm glad you think so.
WEIZSÄCKER:	But they <u>are</u> preventing us from doing any scientific work. I don't see any scientific journals here.
DIEBNER:	And of course no equipment.
GERLACH:	Without journals we can't keep up with the latest research.
HAHN:	Well then, we'll have to stay active by working on topics for which we don't need any journals.
DIEBNER:	Or equipment.
GERLACH:	I agree.
DIEBNER:	I'm not sure the Professor Doctor would agree.
HAHN:	Heisenberg?
DIEBNER:	The great theorist who thinks he knows everything about nuclear engineering—The famous founder of quantum mechanics, the inventor of the uncertainty principle, the boy wonder who became a full professor at age 26 and a Nobel Prize winner at 31. That Heisenberg.

HAHN:	He's definitely the most brilliant physicist I've ever met—except, of course, for Einstein.
DIEBNER:	Three months ago that name would have gotten you killed.
GERLACH:	(*raising his hands thumbs up as he speaks*) Einstein! Einstein! Einstein!
DIEBNER:	(*lowering his hands thumbs down*) Heisenberg! Heisenberg! Heisenberg!
HAHN:	Stop you two! You sound like you've been sipping schnaps.
WEIZSÄCKER:	Or is it heavy water?
GERLACH:	Just the free air of Farm Hall.
DIEBNER:	The point is that you older scientists chose the great quantum genius to speak for German physics during the Reich. And naturally he obtained the most resources for his own reactor project during the war.
HAHN:	Heisenberg was by far the best man for the job, Diebner—And he still is. We needed a successful young physicist to take the political lead. The Nobel Prize certainly indicated success—and international respect.
WEIZSÄCKER:	He was also willing to bear the responsibility that leadership required. They called him a "white Jew" for defending theoretical physics. They said he was promoting "Jewish science" in Germany. They accused him of being homosexual. They tried to put him in a concentration camp.
GERLACH:	Please, Weizsäcker, don't remind me of those dark days.
WEIZSÄCKER:	They put spies in his institute and hidden microphones in his home.
DIEBNER:	But when it came down to the practical work of the project, those of us who were not professors and had to work for the army could not compete, even if we were better at it. And I would have succeeded!
HAHN:	Diebner, we know you tried your best.
DIEBNER:	But because of Heisenberg I couldn't <u>do</u> my best!
WEIZSÄCKER:	And because of you we couldn't do <u>our</u> best!
GERLACH:	As the last Reich nuclear administrator, I tried very hard to prevent that. Even though I wasn't a Nazi, I did everything I could to help my country win the war. Now Germany lies in ruins. And the project didn't even achieve a chain reaction! I failed completely on both counts.
HAHN:	Gerlach, I too love my country and, strange as it may seem, it was for that very reason that I hoped and prayed for her defeat. Even though it meant her destruction, it was the only way to free her from those criminals.
GERLACH:	But you stayed hidden in your lab. (*Upset*) You didn't lose the war!

HAHN:	Gerlach, it's not your fault. You weren't a general. You were a physicist. We scientists couldn't prevent the defeat. And it's not your fault that Heisenberg and Diebner didn't get along.
DIEBNER:	Besides, I almost did succeed with the reactor. Just a little more uranium and heavy water and mine will go critical.
HAHN:	Surely German research is still far ahead of the English and Americans.
GERLACH:	Well…That does help.

Heisenberg appears.

HEISENBERG:	Good afternoon, gentlemen.
DIEBNER:	Heisenberg!

SCENE 4

The scientists are relaxing. Heisenberg is playing a beautiful piece on the piano, then fades as Rittner begins speaking.

RITTNER:	(*to audience*) Major T.H. Rittner to General Leslie Groves, Manhattan Project, Farm Hall Report number 1, July 19, 1945. Top Secret. The arrangements for bringing the party to England went according to plan. The professors and two POW orderlies landed on the afternoon of July 3rd and were taken to the Farm Hall estate near Cambridge by car. Hidden microphones have been installed in all of the rooms and grounds used by the guests. We are transcribing and translating selected conversations which will be forwarded to you periodically starting today. The guests appear to suspect nothing.

He joins the scientists.

RITTNER:	Good afternoon, gentlemen. I would like to welcome you all to Farm Hall. My name is Major Rittner and I have been assigned to your care while you are here.

He shakes hands with each in turn.

HAHN:	Hahn. Good afternoon, Major.
RITTNER:	Good afternoon.
HEISENBERG:	Heisenberg.
RITTNER:	Good afternoon.
GERLACH:	Gerlach.
RITTNER:	Good afternoon.
WEIZSÄCKER:	Weizsäcker. (*Cutting in on Diebner before he can speak*) My goodness, Major. You speak excellent German!
RITTNER:	My family came over from Germany after the Great War. My father has an importing business. I grew up speaking German at home and English at school.

WEIZSÄCKER:	Wunderbar! [*Vunderbar*]
RITTNER:	Well, gentlemen, as you can see we have made every effort to provide for your comfort. Meals are served promptly at 7, 12, and 6. Afternoon tea at 4. I trust that everything is to your satisfaction?
HEISENBERG:	In general, yes. But I object to sharing a room with Diebner.
WEIZSÄCKER:	And I object to sharing with Gerlach. He snores so loudly they could hear him all the way to Belgium!
RITTNER:	Gentlemen, unfortunately we have only a limited number of bedrooms available. So you will have to double up, and—
DIEBNER:	I'll move in with Hahn.
RITTNER:	Oh there you are, Diebner. Good afternoon (*they shake hands*). (*To Hahn*) Is that alright with you?
HAHN:	(*reluctant*) Ah…yes, alright.
HEISENBERG:	I thank you both. I need a room to myself to think and work.
DIEBNER:	(*to Gerlach*) So he can win another Nobel Prize.
HEISENBERG:	Now that that's settled, Major, I would like to ask why you are keeping us here.
RITTNER:	Gentlemen, all I can say is that you are being detained at His Majesty's pleasure.
DIEBNER:	We'll put on a show!
RITTNER:	You are free to move about the designated areas of the house and grounds as you wish, but, gentlemen, I must insist that you give me your word in writing that you will not enter the restricted areas or attempt to escape or to communicate with anyone without my knowledge.
HEISENBERG:	Major, that is asking a lot.
RITTNER:	I know.
HAHN:	But what about our families?
DIEBNER:	I had to leave my wife with those barbarian Moroccan troops in charge!
WEIZSÄCKER:	Can we at least get word to our families that we are alright and learn if they are alright?
RITTNER:	According to my orders you are to communicate with absolutely no one. But I will do everything I can to obtain permission for you to send letters to your families and to receive news from them.
HEISENBERG:	Thank you, Major. I think I can speak for all of us that we will give our word as you request.
HAHN:	Yes.
HEISENBERG:	But surely there's no reason to keep us prisoner here. We have done nothing.

RITTNER:	You're all Nazis, aren't you?
HAHN:	We're nothing of the sort!... Well, not all of us. (*Looks at Diebner*)
RITTNER:	And you were all working for your government on some sort of war contraption.
HEISENBERG:	That was purely a scientific research project.
RITTNER:	I see... (*to all*) Do you mind if I ask a question?
HEISENBERG:	Not at all, Major.
RITTNER:	I was still quite young when I left Germany. I can't imagine what it was like living there during the Third Reich. If you don't mind my asking, how could you stay there for the entire 12 years and work on war research for a government like that if you <u>weren't</u> Nazis? Surely you had many job offers—
HEISENBERG:	That's a difficult question, Major. All I can say is that our goal was to do our best to protect our science and our culture. After all, one doesn't leave a country simply because one doesn't like its leaders.
RITTNER:	You make it sound as if they were just ordinary politicians.
HEISENBERG:	At first they seemed to be.
WEIZSÄCKER:	Then things got progressively worse.
RITTNER:	So why didn't you leave then?
HEISENBERG:	It was a very difficult decision. We tried to protect what we could of German science.
RITTNER:	Yet then you did war research for your government in Berlin.
HEISENBERG:	Major, one has to live under such a regime to understand our situation. All I can say is that we did what we could under the circumstances. We were not politicians. We were physicists. We had no experience with such things.
RITTNER:	I see.
HEISENBERG:	But in a way, we were able to use our research during the war for our own purposes. To save many of our physicists from the front and to prove to the world that decent German physics has survived the destruction. That I am very proud to say.
RITTNER:	But, Heisenberg, what about the people? What about your Jewish colleagues, the non-physicists?...What about Dr. Goudsmit's parents?
HEISENBERG:	Major, please!.... (*upset*) I wanted to help... But perhaps I should have tried harder, maybe tried some other things.... I don't know... In the end we failed. I—I failed. I am very sorry to admit that.... We were so alone, so distrusted. It was just so...... so very difficult.

Pause

WEIZSÄCKER: But now, Major, tell us the <u>real</u> reason why we are here.

RITTNER: I honestly don't know. All I can say is that the Americans consider all of you important enough to keep you here under strict orders.

HEISENBERG: I can understand our importance, but it's all very mysterious. Now they want us to just sit here? And we don't even know when it's going to end!

RITTNER: I wish I could tell you more. I must excuse myself, as I have some duties to attend to. *Exit*

DIEBNER: Heisenberg, why did you promise him we would not try to escape? I'm going to take back my word if I see a chance to escape.

HAHN: But so far there doesn't seem to be a need to escape.

HEISENBERG: Yes, that was my thinking. They are treating us much better than we ever imagined, so we should hold to our word for now. But if we don't hear soon from our families we can threaten to take back our word unless they allow us to contact them.

SCENE 5

Heisenberg writing to Elisabeth, who is silently reading elsewhere on stage.

HEISENBERG: (*speaking as he writes*) My Dear Li [*lee*], I don't know when this letter will reach you, but I want you to know that we are all safe and doing well. An extraordinarily nice Major is caring for us, and I could not imagine a better existence. He has received permission to deliver letters from us and to bring back word from our families.

ELISABETH: (*continues the letter*) Li, I have never felt so fresh and lively in years. There is so much I need to do. I am so full of hope and ambition for the future that for the first time in 12 years I have the feeling that I can really do something worthwhile for Germany. The only thing weighing on me is my concern for you and the children. I have asked the officer who is bringing this letter to you to bring back news of how you are coping.

HEISENBERG: (*continues letter*) I want you to know that when I am finally released from here I will find my way back to you and we will make our lives together as beautiful as they were before all of this happened. So dry your tears and think of me. The worst is now over. We will soon be together again. Your Werner.

ELISABETH: (*savoring the letter*) Your Werner.

SCENE 6

Hahn, Gerlach, and Weizsäcker are sitting around a table playing cards. Diebner is playing a dirge on the piano.

GERLACH: (*throwing his cards on the table*) I've had enough. I can never win at skat.

The others get up from the table.

WEIZSÄCKER: Me, neither.

HAHN: Diebner, would you stop playing that kind of music. It's depressing.

DIEBNER: Why should I? Even if we're living in luxury, it's still a prison.

GERLACH: And we're the enemy prisoners from a defeated nation.

HAHN: Where's Heisenberg?

DIEBNER: I just saw him returning from a run around the grounds….Let's try this…
                 (*Singing while playing a simple melody*)
                 They say His Majesty's pleasure
                 is something we should all treasure.
                 But there is at least one detractor
                 who'd rather work on his reactor
                 and so succeed beyond all measure.

Heisenberg appears, putting on his clothes after a shower.

HEISENBERG: Good afternoon, gentlemen.

DIEBNER: Have a nice run?

HEISENBERG: It's so wonderful to be outside running again in the free air… Sorry to interrupt your recital. Please continue.

DIEBNER: How about this….
                 (*Continues playing and singing*)
                 Yes, here in old Farm Hall
                 he certainly has it all.
                 Fission and fusion
                 there is no confusion
                 and uncertainty reigns over all!

HEISENBERG: (*derisive*) You oughta be in Vaudeville. You're wasting your talents on fission. (*To everyone*) Any news yet on why they're holding us here?

HAHN: Not a word.

HEISENBERG: It's all very strange. They seem to want our knowledge, but they haven't really asked about it.

WEIZSÄCKER:	They could at least let us negotiate our release.
HEISENBERG:	I need to get back to my office. We all need to get back to work.
HAHN:	And to our families! I'm very worried. My wife's in an asylum and my son and grandchild moved to a shelter after we lost our house. Where are they now? How will they survive?
DIEBNER:	It could be they're holding us here while they study our secret documents and discuss them with their experts.
WEIZSÄCKER:	They've probably been using the past two months to imitate our experiments with our equipment and our secret reports.
DIEBNER:	Let's just hope they're the right equipment and reports.
HEISENBERG:	(*angry*) What do you mean?
DIEBNER:	Everyone knows that your reactor design using uranium plates in heavy water was a failure, while my arrangement with uranium cubes nearly succeeded. We don't want them to get the wrong idea about our progress, do we?
HEISENBERG:	Diebner, I worked out the basic theory of reactors and the explosive in late 1939. We got the first evidence of neutron multiplication in '42. There's no doubt that we are now far ahead of the Anglo-Americans.
DIEBNER:	Yet you still insisted on that plate design even though you found my design was much better.

They start shouting.

HEISENBERG:	I had no choice! The order was already at the manufacturer. But the important thing is that my work has put us far ahead of what they now know.
DIEBNER:	And my design has put us far ahead of anything they could have built! If only you hadn't drained our resources I could have succeeded!
HEISENBERG:	If only you hadn't hoarded the heavy water, our last attempt would have gone critical!
GERLACH:	Gentlemen, gentlemen, calm down. We can be sure that once they catch up, they will have to consult with all of us.
DIEBNER:	I hope so.
WEIZSÄCKER:	Doesn't it seem odd that we're the only ones here? What about the others? Why us?
GERLACH:	Because we all worked on the reactor and the explosive during the early years?
HAHN:	Well, I didn't, even though I made the discovery. I didn't want to work on such a thing.
DIEBNER:	But you attended the early meetings on the reactor as the first step to an explosive, and you and your institute remained committed to the project throughout the war.

HAHN:	But only superficially, so we could obtain government funds.
DIEBNER:	So you say.
HEISENBERG:	Gentleman, more importantly, others who did work on fission aren't here. Maybe it's because we're the most important for fission—or for science in general.
WEIZSÄCKER:	But why only us? Did they pick us because one of us is a spy and we wouldn't know it?
GERLACH:	That's nonsense! I know everyone here, and I don't think anyone would spy on the others.
DIEBNER:	Not even for promises of money and freedom?
WEIZSÄCKER:	Maybe you're the spy!
DIEBNER:	Maybe you're the spy!
HEISENBERG:	Oh, stop! I hope nobody would be so foolish.
WEIZSÄCKER:	(*to Gerlach*) But there is someone here whom you don't know.
GERLACH:	Who?
WEIZSÄCKER:	The Major.
GERLACH:	The Major? He doesn't know any science.
WEIZSÄCKER:	Are you sure?
HEISENBERG:	Gentlemen, gentlemen, the war is over. The Third Reich is dead, and we're in "merry old England." All that ugliness is now far behind us. Besides, the Major's hardly here enough to spy on us.
WEIZSÄCKER:	(*looking around*) Well, maybe he doesn't have to be here. Maybe he has secret microphones installed.
HEISENBERG:	(*laughing*) Microphones installed? Oh no, they're not as sophisticated as that. They don't know the real Gestapo methods.
HAHN:	What would they want to find out, Weizsäcker?
WEIZSÄCKER:	What we know about fission. Maybe if we have a nuclear weapon hidden somewhere. They're probably keeping us here until they find out. Then they'll kill us so the Russians can't obtain our knowledge.
HEISENBERG:	I think we're too valuable for that. If they want to find out what we know they can simply ask. I'm willing to tell them everything.
DIEBNER:	If you say so.
HEISENBERG:	Stop saying that.
WEIZSÄCKER:	Well then, why <u>are</u> they holding us here?
HAHN:	Perhaps they don't know what to do with us and are waiting to decide after they have studied our work.
HEISENBERG:	That sounds reasonable.
GERLACH:	When they do decide, I think there's a good chance we'll eventually go back to Germany.

WEIZSÄCKER:	That may be, Gerlach, but it may also be that if we do go back one or more of us will be shot–not by the English but by our own people. They might think of us as deserters or collaborators, and some insane student might decide to shoot us.
GERLACH:	Heisenberg, if we do go back, I hope I can get you to work together with Diebner on completing the reactor.
DIEBNER:	It would be the best chance for a less famous applied physicist to earn a living. The economy is going to be in a mess for a very long time. Perhaps we could even sell the reactor to the Americans.
WEIZSÄCKER:	Or maybe to the Russians.
HEISENBERG:	I think it's more likely they will take us to America to work there on the reactor, and ultimately the explosive. In that case, there is a non-zero probability that we will never see our families again. That would be the worst fate.
HAHN:	That's why all of my hopes and efforts have been directed toward getting in touch with my family.
GERLACH:	Mine too. My wife's all alone in Munich, and the bombing left our house barely livable. There's no food and no money coming in.
HEISENBERG:	Well, my wife and mother and the children are all alone in a small mountain cabin. They can't even grow food in the rocky soil.
HAHN:	You have six children, don't you, Heisenberg?
HEISENBERG:	Yes.
HAHN:	My goodness!
WEIZSÄCKER:	Germany's in chaos. Who's in charge? How can anyone survive?
HAHN:	If only we could send some provisions to our families.
WEIZSÄCKER:	One thing is certain: This can't go on forever.

SCENE 7

Elisabeth writing to Heisenberg, who is silently reading elsewhere on stage.

ELISABETH:	(*speaking as she writes*) My Dear Werner, How glad I am to hear from you at last! I am giving this letter to an American officer here who promised to deliver it to you, wherever you are. I have some very bad news: Your mother has died. She fell and broke her thigh bone several weeks ago. We managed to have her moved to a hospital, but then pneumonia set in. She died there peacefully on July 17.

HEISENBERG: (*breaks down*) If only I could have been there. Now I'm here all alone…. I need you, Li! (*continues the letter*) Werner, I know how close you were to your mother, especially after your father died. I tried to make her as comfortable as possible. She reminded me so much of you, and now I miss you even more. I dream of your return even as I have more sad news to report. It's about the neighbor girl Maria and her son who have been staying with us. Just last week she too died. The doctor could not determine the cause, so we are now all in quarantine. And I am here alone with the children and Maria's son. The boy's father has not been seen since the police took him away because he is non-Aryan.

ELISABETH: (*continues the letter*) Werner, what are we to do? The section of roof you repaired is still leaking. It could collapse with the first heavy snowfall. We need to store wood for the winter. I can't leave the cabin and we're nearly out of food. Can you at least send us some food, even if you yourself cannot come? We all miss you so. The children are constantly asking when you will return. With all my love, your Elisabeth.

SCENE 8

The scientists saunter in, Hahn and Gerlach patting their stomachs.

HAHN: Well, another fine English breakfast! I ate so much I can't eat another thing all day.

GERLACH: Me, too. It's like living on a luxury liner, only we can't get off.

DIEBNER: All we do is eat, sleep, read, and—

WEIZSÄCKER: Do physics!

HAHN: How depressing!

Diebner starts shuffling cards.

WEIZSÄCKER: It's a slow torture for those of us subjected to snoring.

GERLACH: I'm sorry, Weizsäcker. I'm trying to stop. And I don't like your keeping the light on to all hours!

WEIZSÄCKER: I'm just trying to get some work done.

GERLACH: Well, do it during the day.

HEISENBERG: How is it going? Did you make any progress with turbulence?

WEIZSÄCKER: I think I can prove that the onset of turbulence in a fluid flowing in a confined space could help account for the collection of gases around a heavy object in space.

HEISENBERG: That's interesting.

WEIZSÄCKER: Maybe we can connect it with the formation of the solar system or even spiral galaxies.

Diebner makes a particularly loud shuffle.

HAHN:	Would you stop shuffling those cards?
DIEBNER:	Anyone for poker?
WEIZSÄCKER:	I play only bridge and skat.
DIEBNER:	What, too dangerous for you?
WEIZSÄCKER:	Too frivolous.
DIEBNER:	I thought so.
WEIZSÄCKER:	Here, give me those cards. (*He grabs them from Diebner.*)
DIEBNER:	Give them back. (*He grabs them.*)
HAHN:	Would you two stop!
WEIZSÄCKER:	(*To Diebner*) Would <u>you</u> stop!
HAHN:	Please, everyone, let's all try to stay calm. Maybe it would help if the Major could invite some of our English colleagues to visit.
DIEBNER:	That would be a relief.
WEIZSÄCKER:	But apparently no one knows we're here–not even the English scientists.
HEISENBERG:	I wonder if Niels Bohr knows we're here. (*To Weizsäcker*) I still remember our visit with him in Copenhagen four years ago.
WEIZSÄCKER:	Yes, so do I. Do you remember how upset he became when you told him our work revealed the possibility of a nuclear weapon?
HEISENBERG:	I hope by now he's in a better mood. Surely he would not approve of our imprisonment here.
WEIZSÄCKER:	Maybe there is some way we could get a message to him.
HAHN:	Remember your promise to the Major.
HEISENBERG:	Yes, you're right. Let's wait a little longer and see what happens.

SCENE 9

Diebner and Weizsäcker are playing chess alone at a table. They make several moves in silence until Diebner captures one of Weizsäcker's pieces.

DIEBNER:	Ah, ha!
WEIZSÄCKER:	My queen's knight! I should have seen that coming.
DIEBNER:	You should have seen a lot of things coming.
WEIZSÄCKER:	What do you mean?
DIEBNER:	You must have known that Heisenberg was not up to the task.
WEIZSÄCKER:	But he was. He's a great physicist. We didn't have the resources.
DIEBNER:	That's just the point. It doesn't take a quantum genius to see that the resources were not enough to support two competing projects.
WEIZSÄCKER:	Well, if so, then you should have seen the same thing.
DIEBNER:	I had no choice.

WEIZSÄCKER:	Why not?
DIEBNER:	I'm an applied physicist not a physics professor. I don't have an important teaching job and a big research institute to fall back on.
WEIZSÄCKER:	Is that why you worked for the army?
DIEBNER:	So I could pursue nuclear research.
WEIZSÄCKER:	And why you joined the Nazi party?
DIEBNER:	So I could get a job…. But look, Weizsäcker, the war's over. We'll probably go back soon. When we do, why don't we put our differences aside and work together on finishing the reactor? Maybe we could even get Heisenberg to go along. What do you think?
WEIZSÄCKER:	I don't know…
DIEBNER:	Just until we complete the reactor. It would mean a lot to me.
WEIZSÄCKER:	I'm sure it would.
DIEBNER:	But think of what it would mean to you and Heisenberg. A chance to put the past behind you and to guide German science in the right direction. It might even be a bargaining chip to use with the Anglo-Americans.
WEIZSÄCKER:	Most of all it would prove that German science has survived the destruction. That would make it all worthwhile.
DIEBNER:	Yes.
WEIZSÄCKER:	Maybe there is some way we could work together. With your talent and my connections, we should be able to get it running.
DIEBNER:	Even if your father has been arrested.
WEIZSÄCKER:	I'm sure there are people who could get us what we need.
DIEBNER:	Well then…Your move.

He moves.

WEIZSÄCKER:	Check.
DIEBNER:	Damn!

SCENE 10

Possible effect:	*blackout followed by blinding flash of light and a deep rumble As light comes up on stage, an ominous voice is heard.*
VOICE:	Now I am become Death, the destroyer of worlds.

On one side of the stage an upset Hahn is shaking his head in his hands. A nervous Rittner is hovering about. A bottle of brandy is nearby. On the other side, the other scientists are sitting silently around a dining table.

HAHN:	I can't believe it… I just can't believe it!
RITTNER:	I can't believe it either. It was only a brief radio report. The Americans dropped some sort of "atomic bomb" on Japan having the equivalent of 20,000 tons of dynamite! It sounds so fantastic. Could it really be possible?

HAHN:	The awful suffering!…The children!….And I'm the one responsible for it all, the deaths of hundreds of thousands of innocent people!
RITTNER:	Now, now, calm yourself, Professor Hahn. You are not to blame… Here, take this brandy. It will help. (*He pours him a glass and hands it to him.*)
HAHN:	Thank you, Major. (*He chugs it* down)
RITTNER:	Why do you think you're responsible?
HAHN:	If the report is true, then it was my discovery that made it possible.
RITTNER:	What discovery?
HAHN:	Nuclear fission. The splitting of the nucleus. Lise [*Lis-a*] Meitner and I and our assistant Strassmann had been working for several years on bombarding uranium nuclei with neutrons. Then just before the war Strassmann and I found something we never expected!
RITTNER:	What?
HAHN:	The sudden appearance of smaller nuclei in our apparatus. Meitner and her nephew Otto Frisch interpreted our results as the splitting of the uranium nuclei into smaller nuclei. Frisch called it "fission."
RITTNER:	My goodness!
HAHN:	As soon as they calculated the enormous amount of energy released in each splitting of a uranium nucleus, I knew that someday someone would make an explosive stronger than any explosive invented so far. I couldn't sleep for days thinking that I was to blame for bringing the possibility of an "atomic bomb" into the world. And now….my worst fears may be realized…. The terrible suffering! Just thinking about it drove me to contemplate suicide. Perhaps I should have—
RITTNER:	Now, now, Professor Hahn, stop talking nonsense. Just because you made the discovery doesn't mean you are responsible for how it is used.

He hands him another glass. Hahn drinks it.

HAHN:	Perhaps you're right…. I had better tell the others.

Rittner exits as Hahn joins the other scientists. He stands silently at his place.

GERLACH:	Oh there you are, Hahn! We're glad you're in time for the roast beef the Major has provided
HAHN:	Gentlemen, I'm afraid I have some very disturbing news. The Major has just informed me of a radio report a short while ago. According to the report the Americans dropped some sort of "atomic bomb" on Japan early this morning. It had an explosive force equivalent to 20,000 tons of dynamite.

GERLACH:	20,000 tons!
DIEBNER:	That's impossible!
WEIZSÄCKER:	It can't be a fission bomb!
HAHN:	Gentlemen, I can't believe it either. But I am afraid that if it is a fission bomb, then I am to blame.
HEISENBERG:	What? You? Nonsense! You had nothing to do with that! You made a purely scientific discovery. You have no responsibility for how others have used it!
HAHN:	That's what I keep telling myself. (*Hahn slowly sits.*)
HEISENBERG:	If it was a fission bomb, then they were far ahead of us all along.
HAHN:	That would mean you're all second-raters. You might as well pack up.
HEISENBERG:	I quite agree.
HAHN:	They're way ahead of us.
WEIZSÄCKER:	At least now we know why the Americans are holding us here. It was to prevent us from "spilling the beans" about the atomic bomb, as they say.
GERLACH:	Probably also to prevent our capture by the Russians.
DIEBNER:	Goudsmit certainly kept us in the dark.
HEISENBERG:	Yes, he did that very cleverly.
HAHN:	Well, if it is true, how do you think they did it?
HEISENBERG:	They can only have done it if they have isotope separation.
DIEBNER:	You mean uranium 235?
HEISENBERG:	Yes, 235. It's the most fissionable form of uranium.
DIEBNER:	But 235 is so rare that it would require acres of apparatus running for years in order to separate enough of it from the other isotopes.
GERLACH:	That's not absolutely necessary. If they have already built a reactor it would produce plutonium. That would work just as well in a bomb. (*To Hahn*) Do you think they have a reactor running?
HAHN:	I don't believe it.
HEISENBERG:	Did they use the word uranium in connection with the atomic bomb?
HAHN:	No.
HEISENBERG:	Then it's got nothing to do with atoms, but the equivalent of 20,000 tons of high explosive is terrific. All I can suggest is that some dilettante in America who knows very little about it has bluffed them by saying, "If you drop this it has the equivalent of 20,000 tons of high explosive," and in reality it doesn't work at all by fission.
DIEBNER:	Whatever it is, they have a new weapon that we know nothing about.
GERLACH:	Germany lies in defeat, and now German physics lies far behind the Americans.

DIEBNER: At least we still have the reactor.

GERLACH: What is that in comparison?

WEIZSÄCKER: It seems to me that if it really is a fission bomb and it's easy to
                    make, then the Americans know that we will soon find out how
                    to do it if we keep on working. That's a reason to keep us here
                    for a very long time, maybe forever!

DIEBNER: And if we do go back to Germany, they'll prevent us from
                    completing the reactor.

HAHN: I didn't think such a thing would be possible for another
                    20 years. (*To Heisenberg*) Didn't you always tell us we were far
                    ahead of the Anglo-Americans?

HEISENBERG: Hahn, I still don't believe a word about the bomb. I consider it
                    perfectly possible that they have 10 tons of enriched uranium,
                    but not that they have 10 tons of pure U-235.

HAHN: I thought you told us one needed only very little 235 for a bomb.

HEISENBERG: If they enrich the 235 content slightly, they can build a reactor,
                    but with that they can't make an explosive.

HAHN: If they have, let's say, 30 kilograms of pure 235 couldn't they
                    make a bomb with that?

HEISENBERG: It wouldn't go off, because the paths of the neutrons between the
                    nuclei are so large that they can't get a chain reaction going.

HAHN: But then explain to me why you used to tell me that one needed
                    50 kilograms of 235 in order to do anything, and now you say
                    one needs tons!

HEISENBERG: I don't want to commit myself for the moment… But it's
                    certainly a fact that the paths of the neutrons are very long, so
                    one needs a lot of U-235. I'm willing to believe instead that it's a
                    high pressure bomb and that it has nothing to do with uranium.
                    It's a chemical thing.

WEIZSÄCKER: Whatever it is, I think it's dreadful of the Americans to have
                    done it. I think it's madness on their part.

HEISENBERG: You can't say that. You could equally well say, "that's the
                    quickest way of ending the war."

HAHN: That's what consoles me.

GERLACH: Well, then, we'll bet on Heisenberg's suggestion that it's just a
                    bluff.

SCENE 11

The scientists and Rittner are gathered in the living room around a radio.

BBC   (*British accent*) Good evening, ladies and gentlemen. This is the
ANNOUNCER: BBC. Here is the 9 PM news. It's dominated by a tremendous
                achievement of Allied scientists. The greatest destructive power

devised by man went into action this morning over Hiroshima, Japan—the atomic bomb designed for a detonation equal to 20,000 tons of high explosive. British, American, and Canadian scientists have succeeded in using the atomic element uranium for the bomb, while the Germans have failed. At present it's being used for war purposes, but it's expected that further research may make this atomic energy available as a source of power to supplement coal, oil, and hydroelectric plants. President Truman announced that the Allies have spent 500 million pounds and employed up to 125,000 people on what he called the greatest scientific gamble in history—and they've won!... In other news—

Rittner turns off the radio. Chaos reigns. Everyone speaks simultaneously, Rittner outlasting the others.

HEISENBERG:	<u>Then it is true!</u>
DIEBNER:	How could they?!
GERLACH:	We're doomed!
RITTNER:	By Jove, we did it! What an achievement! (*Embarrassed*) I mean.... Well, you know.... Anyway, it looks like we and the Americans have won the race. So I guess the big question now is: What happened? I mean, how did you top scientists lose the race so badly? Surely you knew so much more—
HEISENBERG:	Excuse me, Major, but we did not think of ourselves being in a race to build the atomic bomb. And we nearly succeeded in what we <u>were</u> doing.
RITTNER:	But you worked on uranium for the entire war and now you failed to build even that thing... The...The reactor.
HEISENBERG:	Yes, Major, but we were very close. Just a little more heavy water and it would have worked. If only Diebner and Gerlach had let us have it.
GERLACH:	(*raised voice*) You mean if only you had not insisted on two failed designs!
WEIZSÄCKER:	Oh, let's not start that argument again.
HEISENBERG:	(*angry*) Gerlach, those were only preliminary experimental arrangements!
GERLACH:	But they didn't work! And I had to gain the support of the army and the SS for Diebner's design.
HEISENBERG:	Well, eventually we did shift over to his design, but it was too late.
GERLACH:	Yet you still kept your materials and tried to get even more from Diebner.

All begin to shout.

HEISENBERG: One of us had to move ahead!
WEIZSÄCKER: (*to Diebner*) And <u>you</u> tried to take our reactor away from us!
DIEBNER: But my reactor was nearly critical!
HEISENBERG: Mine was nearly critical! And it was essential that we succeeded before the war ended!
DIEBNER: To make yourself important to the Americans!
HEISENBERG: That's not fair!
RITTNER: Now, now, gentlemen. Calm down.
GERLACH: I've had enough of this! I did the best I could for the sake of Germany. Clearly I failed. And now the Allies have done it, and they've destroyed our cities—and our science in the process. (*Upset*) Germany is ruined!... I am ruined!

He stomps off.

RITTNER: (*following him briefly*) I hope he calms down... (*Returning*) So tell me, what happened with the separation of that rare isotope, uranium 235?
HEISENBERG: We never got our centrifuges up to mass production.
DIEBNER: Even the Americans had produced only a small amount during the year before they went silent.
RITTNER: How did they do that?
DIEBNER: They used what's called a mass spectrograph. A magnetic field separates the paths of moving nuclei of different masses.
WEIZSÄCKER: But that gave them only a small amount of 235–roughly 1 mg.
RITTNER: A tiny fraction of an ounce.
DIEBNER: That's why they must have acres of these devices to make such a bomb.
HEISENBERG: Well, let's try to figure out how many devices. (*Calculating on a small piece of paper*) Let's say each apparatus produces 1 mg of 235 per day. If they set up, say,....one thousand mass spectrographs.... no, one hundred thousand mass spectrographs, that would give them a total of 100 grams of 235 per day. That would be about 30 kilograms of 235 per year and about 90 kilograms after 3 years.
RITTNER: That's about 200 lb.
DIEBNER: You're good!
RITTNER: I've been making such conversions for my family's business all my life.
HAHN: So, Heisenberg. Do you think that would be enough for a bomb?
HEISENBERG: I would think so. But quite honestly I never worked it out as I never believed we could get pure 235.
HAHN: Really?
HEISENBERG: Well, maybe we can figure it out now.

HAHN: It shouldn't be difficult for you, Heisenberg.

HEISENBERG: Let's assume we have a ball of pure U-235 and that each splitting of a 235 nucleus releases two neutrons. Each of those neutrons goes on to split two more 235 nuclei and so on. The number of neutrons multiples very rapidly, producing a chain reaction that goes very fast, creating an explosion.

HAHN: I see.

HEISENBERG: So we can reckon as follows. (*Calculating on an envelope or in a small notebook*) The 20,000 tons of high explosive in the bomb are equivalent to the energy released by about… (*mumbles as he calculates*). 80 generations in a chain reaction. Given the average distance that each neutron travels in 235, and assuming a random walk process, that would require a ball of U-235 with a radius of, (*mumbles as he calculates*)…. 6 times the square root of 80, which is about 54 cm.

RITTNER: That's, ah…. 21 inches.

HEISENBERG: Now plugging in the known density of 235, which may not be quite accurate, that would be……(*mumbles*). about 1000 kg, which is one metric ton!

RITTNER: 1.2 English tons.

HEISENBERG: Just as I thought!

HAHN: But, Heisenberg, given the great difficulty of separating 235 the answer must surely be closer to 100 kg than to 1000 kg of pure 235.

HEISENBERG: (*hesitant*) Yes, I—I suppose you're right.

HAHN: So something must be wrong.

HEISENBERG: But what?

HAHN: I don't know.

HEISENBERG: The measurements are all fairly accurate.

HAHN: There must be something.

HEISENBERG: I don't understand… Maybe it's the amount of energy released per fission…. (*keeps calculating*).

HAHN: Well?

HEISENBERG: (*fumbling around with the paper, very frustrated*) I just can't believe it's not working out!

WEIZSÄCKER: Why not?

HEISENBERG: I don't know! It should be easy enough to do….

WEIZSÄCKER: Certainly.

HEISENBERG: So what's wrong?!… (*keeps calculating*)

WEIZSÄCKER: You of all people should know!

HEISENBERG: (*upset*) This is terrible! It's simply a disgrace that we, the ones who have worked on the problem for years, can't figure out how they did it!

SCENE 12

Serious sounding music on the radio. The scientists are absorbed in reading newspapers.

RITTNER: (*to audience*) To General Leslie Groves, Manhattan Project, Farm Hall Report number 4. On the morning of August 7 our guests read the newspaper reports on the atomic bomb with great interest. Most of the morning was taken up reading these. As the morning progressed, they became more and more agitated.

HAHN: Heisenberg, did you see this? It's from Churchill.

Hahn hands him a newspaper. Heisenberg reads for a few moments.

HEISENBERG: This is outrageous! (*reading*) "The use of the new bomb meant victory in a feverish race with German scientists to find a way to harness and release atomic energy. By God's mercy British and American science outpaced all German efforts. These were on a considerable scale, but far behind. By 1942 it was known that the Germans were working hard to find a way to use such energy to make engines of war with which to enslave the world. The battle of the laboratories went on and has now been won, as have other battles."

GERLACH: We were never in such a race!

WEIZSÄCKER: Because we thought we were far ahead.

HEISENBERG: Apparently so did the Americans, which means they must have considered us the reason for their race to build the bomb. It's all very upsetting.

WEIZSÄCKER: Think about how this will look to those back home. The great German scientists were no match for the Anglo-Americans!

GERLACH: That's right! There must be some way to prevent the newspapers from continuing to make such statements.

HEISENBERG: And to force them to get the facts straight. We weren't even working on a bomb.

DIEBNER: We weren't?

RITTNER: Gentlemen, I have a suggestion. Since you seem so disturbed by these reports, why don't you prepare a memorandum setting forth the details of the work on which you were engaged, and then all of you should sign it.

HEISENBERG: That's an excellent idea, Major–if we can all agree.

DIEBNER: Or agree to disagree.

HEISENBERG: Well then, gentlemen, what should we say in this memorandum?

DIEBNER: That we worked on the uranium reactor and that we made great progress with it.

WEIZSÄCKER: And that we did not work on the atom bomb!

HAHN: But I didn't work on either of those.

DIEBNER:	And I always thought we <u>were</u> working on a bomb! The reactor was just the first step. After all, the army supported your work for several years with that in mind.
HEISENBERG:	Only nominally.
DIEBNER:	Well, what about your report to us at Army Ordnance on a bomb in 1939 and your talk before the education minister in '42 when you described how the new reactor will produce plutonium that could be used as an explosive? Perhaps it would be better if we just said nothing.
HEISENBERG:	But, Diebner, those were only preliminary reports on the theoretical possibilities, and they were intended only to gain the support of the bureaucrats.
DIEBNER:	How did you know they wouldn't order you to build such a bomb?
HEISENBERG:	It was the chance I took.
GERLACH:	Gentlemen, our success must be brought before the world. After all, Hahn's discovery was a German discovery, and our researchers have surely progressed further than the Americans have in building a reactor.
HEISENBERG:	I quite agree.
WEIZSÄCKER:	But we can't say that we tried and failed to make a bomb.
DIEBNER:	Even if we did?
WEIZSÄCKER:	Perhaps we did try at first—
HEISENBERG:	But only in a theoretical sense. The reactor became our primary focus.
DIEBNER:	And if you had gotten the reactor and it began producing plutonium?
HEISENBERG:	Diebner, I don't want to speculate.
DIEBNER:	Well then, can we be absolutely certain that we really have progressed further than the Americans in building the reactor?
GERLACH:	The BBC said so last night.
HEISENBERG:	And today it says here in the *Times* (*reading*): "The problem of controlling the release of nuclear energy has not yet been solved."
WEIZSÄCKER:	Then perhaps they'll need us to build a reactor.
RITTNER:	So, gentlemen, we come back to the question I raised yesterday. Why were the results of your work so meager compared to the Allied achievement? You say that not all of you were pursuing the bomb, but you didn't even have the reactor running. Could it have been because you failed to comprehend the physics?
HEISENBERG:	(*angry*) Absolutely not! Well… even if we didn't fully work out the bomb theory our reactor physics is correct.
DIEBNER:	In theory only!

RITTNER:	Then maybe it was because so many of your best people were forced to emigrate from Germany to the Allies?
HEISENBERG:	That may have contributed somewhat, but I think we still had many excellent people left in Germany. Some of them are here in this room.
RITTNER:	But the Americans targeted only about 40 to 50 of your people for capture! Surely they had many more top people among the 125,000 who built the bomb.
WEIZSÄCKER:	How many people were working with Wernher von Braun on the V-1 and V-2 rockets?
GERLACH:	Thousands worked on that.
HAHN:	Mostly slave labor.
HEISENBERG:	I'm glad we weren't big enough to need slave labor.
HAHN:	But wasn't slave labor used in the uranium production?
HEISENBERG:	I don't know…. Maybe it was.
GERLACH:	One thing is clear: We were not able to work on the scale of the American effort.
DIEBNER:	Nor could we achieve large-scale cooperation. There were at least 9 different research groups, and each group said the others were incompetent. Even the German post office had its own project!
GERLACH:	(laughs) I tried to stop them.
DIEBNER:	Who wants to receive a radioactive letter?
WEIZSÄCKER:	Unfortunately, Gerlach, you failed.
GERLACH:	Now don't get me started!
WEIZSÄCKER:	Think of it this way. If we had worked together on a larger scale to achieve the bomb and then failed, the German authorities would have killed us. But if we had succeeded and they dropped the bomb on London, then the English authorities would have killed us as war criminals. Let's just be grateful we're still alive.
DIEBNER:	If you haven't got the courage, then it's better to give up at the start!
GERLACH:	Don't always make such aggressive remarks! As the head administrator I tried to keep both reactor projects going in the hope that at least one of you would succeed.
DIEBNER:	Obviously neither one did, but mine was better, and the Americans were better than all of us!
GERLACH:	Would you stop trying to contradict me!
HEISENBERG:	And me too! (pause) All I can say is we wouldn't have had the courage to recommend to the government in the spring of 1942 that they should employ 125,000 men just for building a bomb that might not work.

WEIZSÄCKER: (*emphatically*) I believe the reason we didn't do it was because all the physicists didn't want to do it, on principle. If we had all wanted Germany to win the war, we would have succeeded!

HAHN: I don't believe that, but I am thankful we didn't succeed.

HEISENBERG: The point is that the whole structure of the relationship between the scientist and the state in Germany was such that although we were not 100% anxious to do it, on the other hand we were so little trusted by the state that even if we had wanted to do it, we would have had a difficult time getting it approved.

DIEBNER: Because the officials were only interested in immediate results, like rockets and jet airplanes. They didn't want to work on a long-term policy as the Americans did.

HAHN: Even if you had gotten everything you wanted, it is by no means certain that you would have gotten as far as the Americans and English have now. After all, they were bombing our cities and factories.

DIEBNER: And we were squandering our resources, stumbling over our theories, and failing to work together.

WEIZSÄCKER: Diebner, there is no question that we were very nearly as far as they were, but it is a fact that we were all convinced that the thing could not have been completed during the war.

HEISENBERG: Well, that's not quite right. I would say that I was absolutely convinced of the possibility of our making a reactor, but I never thought we would make a bomb, and at the bottom of my heart I was glad that it was to be a reactor and not a bomb. I must admit that.

DIEBNER: And if it had been a bomb and not a reactor?

HEISENBERG: Ah.... Diebner, I told you before, I refuse to speculate about such a thing.

WEIZSÄCKER: Gentlemen, I don't think we ought to make excuses now because we did not succeed, but we must insist that we did not want to succeed.

HAHN: Please, Weizsäcker, let's stay with the facts.

DIEBNER: I think it's characteristic that the Germans made the discovery and didn't use it, whereas the Americans have used it.

WEIZSÄCKER: In fact, one could say that the peaceful development of the uranium reactor was made in Germany under the Hitler regime, whereas the Americans and the English developed this ghastly weapon of war.

HEISENBERG: I would object to putting that in our memorandum, Weizsäcker. We already have enough trouble explaining our position to the Anglo-Americans.

GERLACH: Yes, but on the other hand we have to explain to our countrymen why we did not achieve a bomb. You can be certain there are many people in Germany who will say that it is our fault that Germany lost the war. And they would be right.

DIEBNER: When we get back to Germany we will be looked upon as the ones who destroyed everything. We won't remain alive there for long.

GERLACH: Clearly the wording of this memorandum will have to be done very carefully. Our reputations as scientists and as men are at stake on both sides, with the Germans and the Anglo-Americans.

WEIZSÄCKER: If this memorandum doesn't go right, we could be shunned or even killed by either side, even by both sides!

HAHN: Gentlemen, I'm sorry I can't participate. I was not involved in your research or in the many arguments you are presenting now for your lack of success.

HEISENBERG: But you were involved in the discovery. And for that you should participate. For the rest of us we cannot afford to have divergent explanations even though some of us were working in divergent directions.

RITTNER: (*to audience*) There was considerable discussion among the guests about the wording of the memorandum. They did not complete the task until the following day.

Each reads to the audience in turn.

HEISENBERG: "Memorandum. August 8, 1945. As the press reports during the last few days contain partly incorrect statements regarding the alleged work carried out in Germany on the atomic bomb, we would like to set out briefly the development of our work on the uranium problem.

HAHN: 1. The fission of the atomic nucleus in uranium was discovered by Hahn and Strassmann in Berlin in December 1938. It was the result of pure scientific research which had nothing to do with practical uses. It was only after publication that it was discovered almost simultaneously in various countries that it made possible a chain reaction of the atomic nuclei and therefore for the first time a technical exploitation of nuclear energy.

DIEBNER: 2. At the beginning of the war a group of research workers was formed in Germany with instructions to investigate the practical application of nuclear energy.

GERLACH: Towards the end of 1941 the preliminary scientific work had shown that it would be possible to use nuclear energy for the production of heat and therefore to drive machinery.

WEIZSÄCKER: On the other hand, it did not appear feasible at the time to produce a bomb with the technical possibilities available in Germany.

HEISENBERG: Therefore, the subsequent work was concentrated on the problem of the reactor for which, apart from uranium, heavy water is necessary."

RITTNER: (*to audience*) All of the guests signed the document and formally presented it to me in a small ceremony.

Scientists in the background signing the document.

On the next day, August 9th, the BBC announced that the United States had dropped a second atomic bomb, this one on the Japanese city of Nagasaki. The scientists were staggered to learn that the bomb was powered by plutonium produced by several large reactors that had been running for over a year in the United States.......

Scientists in background scatter off stage.

(*Slowly*) That evening an eerie silence settled over Farm Hall.

SCENE 13

Hahn is at a window looking out. The other scientists are sitting around the living room reading newspapers and sipping tea.

HAHN: Look out there. All of those beautiful trees are practically bare. The autumn wind has blown their leaves away. Only the bare branches remain, exposed against the cold grey sky.

RITTNER: (*to audience*) Farm Hall Report No. 18. During the nearly 10 weeks since the news of the atomic bombs my superiors finally informed me that our guests are being held so we can determine how far they progressed in building the atomic bomb, to keep the prospect of such a bomb absolutely secret until it was used, and to prevent their capture by the French or Russians. Meanwhile, the guests' concerns about the future have been alleviated somewhat by discussions with British scientists concerning their eventual release. On October 26, 1945 the scientists received some more startling news.

HEISENBERG: Hahn, read this! (*He tries to hand him the newspaper.*)

HAHN: No, No. I don't have time for that right now.

HEISENBERG: But Hahn, this is very important for you. It says here that you have won the Nobel Prize in Chemistry.

Speaking over each other.

DIEBNER: The Nobel Prize!

GERLACH: What!

WEIZSÄCKER: My goodness!

HAHN: I don't believe it. It's some kind of trick.

Everyone tries to congratulate him.

HAHN: No, no, no. This is absurd. I can't imagine the Nobel Prize going for such a thing.

HEISENBERG: But, Hahn, you're the one who discovered nuclear fission. And for the benefits it will bring to mankind you have been selected to receive the Nobel Prize.

HAHN: So far all it has brought mankind is suffering and destruction.

WEIZSÄCKER: How ironic that the man who invented dynamite used the money to create a prize for the improvement of mankind that is now awarded to the man who invented the atomic bomb!

GERLACH: That's ridiculous! He didn't invent the atomic bomb.

HAHN: Thank you, Gerlach, but Weizsäcker is right. It's entirely inappropriate.

HEISENBERG: Well, then, if I can't congratulate you for the prize, can I at least congratulate you for the 6000 lb you are to receive?

GERLACH: (*looking at newspaper*) It says here that you were actually awarded the prize last year, but the announcement was withheld because the German government forbade its citizens to accept Nobel prizes.

DIEBNER: Does it say anything there about Meitner and Strassmann? After all, Strassmann worked with Hahn on the observation, and Meitner and her nephew Otto Frisch worked out the theoretical explanation.

GERLACH: (*scanning paper*) I don't see anything….

HAHN: Why should it mention them? Meitner left several months before the discovery, and Strassmann was only an assistant.

WEIZSÄCKER: In fact, it could be said that Meitner might have impeded the discovery if she had stayed.

HAHN: No! That's absolutely wrong! She helped us with the work in Berlin almost to the end. But she had to leave when Germany annexed Austria, because, even though she converted, she is of Jewish descent. After she settled in Sweden she and her nephew used Bohr's theory to explain to us that what we were observing was fission of the uranium nuclei. Never in my wildest thoughts did I imagine that a nucleus could be split in two.

RITTNER: Then maybe she too should get the Nobel Prize?

HAHN: (*briefly reflects.*) I think we should leave that for the committee to decide.

RITTNER: Gentlemen, I can inform you that we have just received initial word from our London office that this report is true. And in anticipation of the final news, we are already planning a small celebration this evening with some fine red wine provided by my superiors. We are truly honored now to have 2 Nobel Prize winners among our guests: Professors Heisenberg and Hahn.

HEISENBERG:	(*lifts a tea cup*) Here's to Otto Hahn!
ALL:	To Otto Hahn!
DIEBNER:	(*alone*) And to Lise Meitner! And to Fritz Strassmann! And to Otto Frisch!

SCENE 14

Heisenberg writing a letter to Elisabeth.

HEISENBERG:	(*speaking as he writes*) My Dear Li, the events of the past weeks have shaken us all. Hahn has received the Nobel Prize, but he is still devastated by how his discovery has been used. I know many of the English and American colleagues who worked on the bomb. Some were my students. I feel very sorry for them, because their names are now associated with this horrible thing. Perhaps this will awaken a new feeling of togetherness among all people in such a dangerous world.
ELISABETH:	(*continues the letter*) Li, finally some good news. It looks like we will be coming home soon. Now I long even more to be with you again in all peace and quiet, and sharing with you all the love that is in my heart.
HEISENBERG:	Please send me some pictures of the children. I can't wait to see them. Tell me all about your plans for Christmas. I hope I can celebrate it together with you and the children as we always have. With love, Your Werner.

Elisabeth treasures the letter.

SCENE 15

Rittner on one side of the stage.

| RITTNER: | (*to audience*) In November 1945 I traveled to a meeting of British and American officers in Brussels regarding the future of our guests at Farm Hall. I left my executive officer in command at Farm Hall. At Brussels we agreed that the scientists will be returned to northern Germany on January 3rd, 1946 and released within the British occupation zone. Their families will be brought to them there. After the meeting Dr. Goudsmit invited me to accompany him on his first return visit to his parents' home in The Hague. |

The Goudsmit home. Goudsmit appears.

| GOUDSMIT: | Here it is, Major. This is the house... (*enters*) I can't believe it! |

Rittner follows Goudsmit into the house.

| RITTNER: | My goodness. What a mess! Even the door jamb is ripped out. |

GOUDSMIT: You wanted to know what happened after they left.

RITTNER: I never imagined it would be like this!

GOUDSMIT: Neither did I. The neighbors stripped it of everything that will burn. It was probably to keep warm during the winter. They knew my parents were not coming back.

RITTNER: Even the staircase is gone.

GOUDSMIT: (*Walking around*) Here was the glassed-in porch that was my mother's favorite breakfast nook. That's where the piano once stood. Over here is where my bookcase has been since I can remember. What happened to all of my books?

RITTNER: Probably burned along with everything else.

GOUDSMIT: And look at the garden out there. All of the flowers are gone. Everything is dead... Only the old lilac bush is still standing. (*He looks up with his eyes closed*) When I close my eyes I can still see the house as it was earlier. I can even feel the presence of my parents at home waiting for my return just as before.... (*shouts painfully*) If only I had been here! If only I had gotten the American visas to them in time! Everything was ready for their escape...

RITTNER: Dr. Goudsmit, you did everything you could.

GOUDSMIT: But it wasn't enough. I could have done more....much more.

RITTNER: Dr. Goudsmit, what is this room over here?

GOUDSMIT: That's my room. (*He picks up some papers from the floor.*) Look, these are my old high-school report cards! My parents saved them all these years. I can't believe they survived.

Rittner picks up something.

RITTNER: And what's this? A framed photograph. The glass is broken.

Goudsmit comes over and looks at it.

GOUDSMIT: My goodness! It's Heisenberg with me in the garden of this house! There's the lilac bush in full bloom. My father took this picture when Heisenberg visited us in 1925. I had it on my desk when I left. I can't believe it's still here.

RITTNER: What a coincidence!

GOUDSMIT: What makes it more so is that vague letter he wrote after my father and mother were arrested in this very house.

RITTNER: And now to have the house destroyed like this....

GOUDSMIT: (*very bitter*) If only **he** had done more!

RITTNER: I'm sure those were very difficult times.

GOUDSMIT: He's such a great physicist. He made so many fundamental breakthroughs in quantum mechanics, nuclear physics, particle physics.... How could he think a mere letter would suffice?

RITTNER: Maybe we expect people who are great in one area to be great in other areas as well.

GOUDSMIT: I suppose so….
RITTNER: Not just in research but in life itself.
GOUDSMIT: (*hesitantly*) Maybe, we… we even expect it of ourselves.
RITTNER: That is the most difficult.

SCENE 16

RITTNER: (*to audience*) On January 3rd, 1946, all of the guests assembled at
              the nearby airbase to board a plane for their flight back to Germany.
              Dr. Goudsmit and I, along with a military guard for their protection,
              joined them on their journey to the beginning of their future lives in
              the new postwar Germany.

*As their names are called, the characters cross the stage as if crossing a tarmac to
the plane, pausing half way as Rittner speaks, before exiting on the opposite side.*

RITTNER: Kurt Diebner…. After Diebner returned to Germany he founded a
              company in Hamburg for the commercialization of nuclear energy.
              He worked there on the design of nuclear powered ships.
RITTNER: Carl Friedrich von Weizsäcker….Weizsäcker became a philosophy
              professor at the University of Hamburg. In 1957 he rallied the other
              guests—except Diebner—for a new memorandum that successfully
              opposed a NATO plan to equip the West German army with nuclear
              weapons.
RITTNER: Walther Gerlach…. Gerlach returned to Munich where he became
              rector of the University of Munich.
RITTNER: Otto Hahn… Hahn returned to West Berlin and served as president
              of the Max Planck Society, a network of government-sponsored
              research institutes.
RITTNER: Werner Heisenberg….Heisenberg headed the Max Planck Institute
              for Physics in Göttingen and Munich. He was influential in the
              reconstruction of West German science and in the development of
              West German science policy.

He waits as Elisabeth enters. They embrace, happy to be together again.

RITTNER: Elisabeth Heisenberg….Elisabeth gave birth to her seventh child in
              1950. One of the boys became a physicist. Another became a
              professor of microbiology. Her oldest daughter is a psychologist.

They exit together.

RITTNER: Samuel A. Goudsmit… Dr. Goudsmit became a senior scientist and
              chairman of the physics department at the Brookhaven National
              Laboratory on Long Island, New York. He also served for many
              years as editor-in-chief of *Physical Review* and as the founding
              editor of *Physical Review Letters*.

RITTNER: Major T.H. Rittner…. After a brief tour in Germany with the British occupation I left the army and re-joined my family's importing business. The 153 pages of my Farm Hall reports were released to the public in 1992. The Farm Hall estate was returned to private ownership soon after the war. It is still there today.

He follows the others off stage. Lights fade on empty stage.

Elisabeth and Werner Heisenberg soon after the war (Max Planck Society, courtesy of AIP ESVA)

A Brief History of the German Project, Alsos, and Farm Hall

During the German scientists' many discussions at Farm Hall about their lack of progress toward an atomic bomb in comparison to the Allied success, Heisenberg made the following observation.

> The point is that the whole structure of the relationship between the scientist and the state in Germany was such that although we were not 100% anxious to do it, on the other hand we were so little trusted by the state that even if we had wanted to do it, it would not have been easy to get it through.

That distrustful relationship not only hindered the scientists' work, but it also did encourage their willingness to work on nuclear energy for the state in order to demonstrate the value of themselves and their science to the regime and thereby achieve certain benefits. Among these were the silencing of ideological attacks and threats, the exemption of scientists from the front to work in the laboratory, and the restoration of scientists' influence on education and some policy decisions.

The relationship between German science, especially physics, and the state gradually evolved from the very beginning of the Third Reich. The following account derives from my previous work (esp. Cassidy 2001, 2009) and from that of many others who have written on this subject (esp. Walker 1989, 1995; Hoffmann and Walker 2007; Beyerchen 1977; Ball 2014). We begin with the pre-war years of the regime, because it was during those seven years of the twelve-year Reich that the German scientists made their decisions about staying in Germany and about their personal stances toward the regime and their continued work under it. This "set the stage" for their responses to the discovery of fission and the outbreak of war in 1939. It is unfortunate that the pre-war years are often overlooked by those attempting to comprehend the scientists' behavior during the war years; the earlier years are essential for a complete understanding of the scientists' motives and behavior during the later years.

© Springer International Publishing AG 2017
D.C. Cassidy, *Farm Hall and the German Atomic Project of World War II*,
DOI 10.1007/978-3-319-59578-8_2

The Pre-war Regime

On January 30, 1933, Germany's president Field Marshall Paul von Hindenburg, appointed Adolf Hitler, chairman of the National Socialist German Workers Party —derisively known as the Nazi Party—parliamentary chancellor and head of a new cabinet in Berlin. Although Hitler promised to return Germany to its pre-world war greatness, from the very moment Hitler and his henchmen came to power Germany descended inexorably deeper into dictatorship and depravity, ultimately leading to the horrors of world war and the Holocaust.

The nightmare prospect that Hitler could get his hands on the atomic bomb drove the United States and its western Allies to do everything they could to achieve the bomb first. And they succeeded by mid-1945. But by then the bomb was no longer needed to end the European war. A defeated Germany had surrendered on May 8, 1945. Twelve years after Hitler's rise to power many millions of people had died, most of Europe was in ruins; and the once mighty Germany now lay flat on its back, devastated by round-the-clock bombing, its economy and infrastructure in collapse, its people in shock, and its territory divided among the four Allied occupation powers. Its atomic scientists had not even achieved a self-sustaining nuclear reactor.

One month after his appointment as chancellor, Hitler, responding to the burning of the Reichstag (parliament) building, issued an emergency decree suspending the constitution. By summer, thousands of Jewish civil servants had lost their jobs, and many were leaving the country. The first concentration camps were already filling with political opponents, among them pacifist resistors and opposing clergy. Although educators, particularly professors, as representatives of German culture, enjoyed a very high social and cultural standing in Germany, they, too, were hard hit by the dismissal policy. With no private schools or universities in Germany, all teachers and professors were civil servants subject to the anti-Semitic laws, which were enforced by the cultural ministry of each state

The sudden dismissals of leading Jewish physicists, the resignations of those temporarily spared because of war service at the front, and the student harassment of those who stayed (students were on the vanguard of the Nazi movement in the universities), demanded responses from leading non-Jewish physicists, despite their years of "apolitical" withdrawal. Like most educated Germans they believed that Hitler's "excesses" would soon subside under the pressure of governing, and that eventually the crude "ruffians" would be voted out of office. Max Planck, the 75-year-old de facto "dean" of German physics, counseled quiet diplomacy. After all, he reasoned, professors still commanded great respect (Heilbron 1986). Public resistance or street demonstrations would lead only to violent suppression by Hitler's storm troopers. Thus in April 1933 Planck went directly to Hitler, the German Führer (leader), in an attempt to convince him of the damage he was doing to German science, until then the world's leader in many scientific areas. The dictator flew into such an anti-Semitic rage that the aged quantum theorist could do nothing but leave in dismay. When the 31-year-old Heisenberg came to Planck seeking guidance soon afterward, a disheartened Planck advised him to hunker down until the storm blew over—assuming it would blow over—and to maintain in the meantime "an island of existence" in his university, Leipzig, in which decent German physics could survive until the catastrophe had passed. Heisenberg took Planck's advice, and, as one of the most prominent physicists remaining in Germany, he began to identify his personal survival in Germany with the survival of decent German physics.

At the end of 1933 Heisenberg received the Nobel Prize for physics as well as the prestigious Max Planck Medal of the German Physical Society. A new international star of German physics was born. As the regime's assaults on the profession continued, the profession's leaders, Planck and Max von Laue, selected the youthful Heisenberg to lead the profession's new response, even though he was not particularly adept in such matters. Heisenberg accepted the task. With further guidance from his friend and former student Carl Friedrich von Weizsäcker, whose father was a diplomat in the Reich's Foreign Ministry, Heisenberg pursued a diplomatic strategy during the next two years. He helped launch a petition campaign against regime policies affecting the profession; he held personal meetings with state bureaucrats in an effort to rescind individual dismissals; and he attempted to convince Jewish physicists spared during the first round of dismissals to remain at their posts. When they resigned or were eventually dismissed, he worked to find suitable replacements. But by the end of 1935 all of these efforts, some of questionable ethics and efficacy, had failed. The German physics profession had lost many of its physicists, among them many leading figures. Heisenberg, like many others, retreated into his island of existence—into the small circle of his students, his physics, and his music. He was by then a concert-level pianist.

But the Third Reich's descent into terror proceeded unabated. The battle in education and academic research now turned to the control of the science, mathematics, and technology faculties, and ultimately to control over what they would teach. Professional organizations, such as the German Physical Society, came under strong pressure for "self-coordination," voluntary alignment with Nazi policies including the exclusion of Jews from membership (Hoffmann and Walker 2007). Nazi ideology became the weapon of choice for those lusting for influence in these professions. In mathematics and in most of the sciences "German" ideologies emerged, pitting "German" (*deutsche*) or "Aryan" science against "Jewish science." In physics, the Nobel Prize winning experimentalists Philipp Lenard and Johannes Stark championed the supremacy of "Aryan physics"—essentially 19th century "classical," experimental physics—in teaching and research; and they demanded the suppression of "Jewish physics," which they identified as the non-classical fields of relativity theory and quantum mechanics, both often associated with Albert Einstein.

The conflict came to a head in 1935 when a faculty search committee at the University of Munich, headed by quantum experimentalist Walther Gerlach, placed Werner Heisenberg at the top of its list to succeed the great theoretical physicist Arnold Sommerfeld. Sommerfeld, one of the founders of quantum theory, had retired from the prestigious Munich chair for theoretical physics in that year. Since teaching appointments were made by state education officials upon the recommendation of the university faculty, the Aryan physicists leapt into action. In January 1936 the official Nazi party newspaper, with a circulation of millions of party members, published an article by a physics student calling for the replacement of "Jewish physics" by "Aryan physics" at all German universities.[1] It was a clear challenge to Heisenberg's appointment and to all subsequent appointments of non-Nazi physicists representing contemporary physics. Backed by his colleagues, Heisenberg took up the challenge. He managed to publish a refutation of the article in the same party newspaper (Heisenberg 1936), while at the same time he pressured Reich education officials by circulating, together with physicists Hans Geiger and Max Wien, a letter of protest that was signed by nearly the entire physics profession. These efforts, together with the diplomacy of Leipzig officials within the higher bureaucracy, left the matter in a stalemate.

A year later Gerlach and the Munich faculty once again named Heisenberg as their top choice to succeed Sommerfeld. But by that time Nazi demagogues were taking another step deeper into depravity. Having driven most of the Jews from public life, including education, the Reich leaders unleashed a broad campaign of intimidation and suppression of any further overt or covert opposition to Nazi policies among non-Jews throughout the German populace. In their parlance, such opponents were called "white Jews," making them susceptible to the same treatment as Jewish victims. With Heisenberg once again in line to succeed Sommerfeld, Stark seized the opportunity to silence the opposition among physicists by placing (if not himself writing) an article in the newspaper for SS functionaries, *The Black Corps*, under the title "'White Jews' in Science."[2] The article singled out Planck, Sommerfeld, and especially Heisenberg as "white Jews" who "must be eliminated just as the Jews themselves." A large facsimile of Stark's signature appeared on the page.

Stark's attack appeared in the SS newspaper in July 1937, just as Heisenberg arrived in Munich to take up residence along with his new bride, the former Elisabeth Schumacher, the daughter of a Berlin global economics professor. The future of the German physics profession, and of Heisenberg himself, were now at stake, but the time for petitions and personal lobbying was over. There was only one thing to do: attempt to reach the head of the SS himself, Reichsführer-SS Heinrich Himmler, and somehow persuade him to denounce this vile attack on Heisenberg. Otherwise, Heisenberg realized, he too would likely have to leave Germany or face constant insults and threats if he stayed, but leaving was a step that he, like most other Germans, could not bear to contemplate. "You know that it would be very painful for me to leave Germany," he wrote Sommerfeld. "I do not

---

[1] Willi Menzel, "Deutsche Physik und jüdische Physik," *Völkischer Beobachter 49*, no. 29 (29 January 1936), 7.

[2] "'Weisse Juden' in der Wissenschaft," *Das Schwarze Korps* (15 July 1937): 6, English translation in Hentschel (1996, 152–156).

want to do it unless it must be absolutely so. However, I also have no desire to live here as a second-class person."[3]

Fortunately, Heisenberg had a personal connection to Himmler. Himmler's father and Heisenberg's grandfather, a Gymnasium rector, had been on the Bavarian school board together in earlier years. In this way Heisenberg's mother had become acquainted with Himmler's mother. The former managed to convince the latter to relay a personal letter from Heisenberg to Himmler. In the letter Heisenberg requested a decision from the Reichsführer on whether or not he approved of Stark's attack. If he approved, then Heisenberg would resign his professorship and presumably accept one of the many job offers he received from abroad. But if Himmler disapproved, Heisenberg demanded the restitution of his honor and protection from further attacks. It was a bold move, but nothing short of that was needed.

Heisenberg's letter led to a year-long intensive SS investigation. Secret microphones were placed in Heisenberg's home and office, spies were placed in his classroom and seminars, and the SS subjected him to night-long interrogations in the infamous basement chambers at Gestapo headquarters. The physics professor somehow survived the frightful, humiliating scrutiny. With the help of an SS physics doctoral student and key recommendations in Heisenberg's favor, exactly one year later, in July 1938, Himmler wrote a personal letter to Heisenberg disapproving of Stark's attack and assuring him that he would prevent any further assaults.[4] Heisenberg would remain in Germany and continue to preserve decent German physics. But, despite Gerlach's efforts, Sommerfeld's chair went instead to an unqualified Nazi applied physicist. An enraged Gerlach declared physics dead in Munich. Himmler promised Heisenberg an appointment to another prestigious position as proof of his "exoneration."[5]

Five months after Heisenberg received Himmler's personal protection, Otto Hahn, a radiochemist at the state-sponsored Kaiser Wilhelm Institute for Chemistry in Berlin, observed a very strange by-product emerging from the bombardment of uranium atoms by neutrons.

Nuclear Fission

In December 1938, Otto Hahn reported to his long-time colleague at the Berlin Institute for Chemistry, Lise Meitner, the latest results of his uranium experiment, performed together with his assistant, Fritz Strassmann. Meitner had recently fled the Reich after Hitler annexed her homeland, Austria. Meitner, of Jewish heritage,

---

[3]Heisenberg to Sommerfeld, 14 April 1938, Sommerfeld Papers, Deutsches Museum, Munich, and Archive for History of Quantum Physics (AIP NBLA and elsewhere), microfilm 31, Sect. 5.

[4]Published in Goudsmit (1996, 117–119).

[5]Heisenberg to Sommerfeld, 13 May 1939, Sommerfeld Papers, note 3.

could no longer receive protection as a foreigner. She had recently landed in Stockholm, Sweden at a new research laboratory sponsored by the Royal Swedish Academy of Sciences, the academy that awarded the Nobel Prizes (Sime 1996).

Several years after the discovery of the neutron in 1932, Meitner, Hahn and Strassmann had begun sending neutrons onto uranium, element 92, the last element on the periodic table, in the hope of producing the unknown next element, element 93. (The numbers 92 and 93 refer to the number of protons in each nucleus.) As suggested by Italian physicist Enrico Fermi, a uranium nucleus with 146 neutrons, known as isotope U-238, would absorb the incoming neutron and become U-239 (239 refers to the total number of protons and neutrons in the nucleus.). The absorption would induce one neutron in the nucleus to decay into a proton with the ejection of an electron (and, as later confirmed, an anti-neutrino), a process known as beta decay. With the new proton, the nucleus should now be the hitherto unknown element 93, which today we call neptunium (after the next planet beyond Uranus).

A perplexed Hahn wrote to the exiled Meitner that in some of their experiments he was not observing element 93 but element 56, barium, half across the periodic table! He pleaded with Meitner, the physicist of the group and adept also in theoretical work, to solve this conundrum. Just at that moment, Meitner's nephew, Otto Robert Frisch, also a physicist and Jewish refugee, had come from Bohr's institute in Copenhagen to visit his aunt in Sweden. Using Bohr's liquid-drop model of the nucleus, Meitner and Frisch discovered that, in fact, the uranium nucleus had split, what Frisch called "fission," after the reproductive splitting of a biological cell. One half would be a barium nucleus, while the other half of the split nucleus should be element 36, krypton, since 56 + 36 = 92. They also predicted that each fission of uranium should release an enormous amount of energy.

Hahn and Strassmann quickly published their still puzzling observations—but without inviting Meitner to join them or even mentioning her collaboration. Hahn later said he was forbidden to publish a paper with a Jewish co-author. Meitner wrote back to Hahn informing him of their fission theory, which gave Hahn the confidence to report that the nucleus had indeed split. Shortly before Frisch returned to Copenhagen to confirm the result experimentally, he and Meitner reported their theory and its confirmation in a paper published in a British journal in January 1939, just after Hahn's and Strassmann's paper appeared in a German journal.

Five years later, in 1944, the Swedish Academy awarded Hahn alone the Nobel Prize in Chemistry "for his discovery of the fission of heavy nuclei." Meitner, Frisch, and Strassmann went unmentioned. Hahn made little effort to correct the oversight. However, the Nobel committee delayed announcing the award of the prize to Hahn until after the end of the war in 1945, when Hahn was then at Farm Hall. Hitler had forbidden Germans to accept the prize after the Nobel Peace Prize for 1936 went to Carl von Ossietzky, a German pacifist who was at that moment being beaten and starved to death in the Dachau concentration camp near Munich. In 1944, no one yet knew of the awesome, destructive power that nuclear fission would unleash.

Meanwhile, upon his return to Copenhagen, Frisch had immediately informed Bohr of the discovery of fission just before Bohr left for an extended stay in the

United States. After bringing the news of fission to America, Bohr and physicist John Wheeler at Princeton University worked out, and even published for all to see, the first extensive theory of nuclear fission, including both the potential controlled release of nuclear energy in a reactor and its uncontrolled release as an explosive.

Bohr and Wheeler found that in a fission chain reaction the fission occurs primarily, not in the plentiful uranium isotope U-238, but in the extremely rare isotope U-235. (U-238 constitutes about 99% of natural uranium; U-235 only about 0.7%.) Each fission splits the U-235 nucleus with the release of a great amount of energy, along with 2 to 3 neutrons. These high-speed neutrons can then go on to split more U-235 nuclei if they can find them nearby. When enough U-235 is present to enable this, a "critical mass" extracted from naturally occurring uranium, then a chain reaction will occur with the release of an unparalleled amount of energy within a fraction of a second—an atomic bomb (although it was not yet certain that it would be small enough to be loaded as a bomb on an airplane). The great difficulty, however, was the extraction of a critical mass of the rare U-235 from natural ore obtained from uranium mines. For separating out U-235, the Manhattan Project employed highly sophisticated centrifuges and diffusion techniques on an unparalleled industrial scale. At Farm Hall and in the play, Heisenberg's calculation of the scale of the effort required astounded the scientists. Some countries ever since have been using similar techniques in their attempts to acquire nuclear weapons. The atomic bomb dropped on Hiroshima, Japan on August 6, 1945 was powered by a critical mass of extracted U-235, resulting in an explosion equivalent to the detonation of nearly 20,000 tons of TNT (see Rhodes 1986).

An alternative route to the atomic bomb lay through the reactor. Bohr realized that when a high-speed neutron strikes a U-238 nucleus it will either fission the nucleus or be absorbed by it. On the other hand, a slow moving neutron bounces off U-238, but it always fissions a U-235 nucleus. Thus, an energy producing reactor, in which the chain reaction is controlled, can be constructed using natural uranium, or uranium slightly enriched in U-235, by introducing a material called a moderator to slow but not absorb the neutrons produced by fission. The slowed neutrons are then sent back into the uranium to fission more U-235 but without creating a full chain reaction. The Germans used heavy water alone as a moderator, while the Manhattan Project used both heavy water and ultra-pure graphite. The German reliance on heavy water greatly hindered their progress toward a self-sustained reactor when the project encountered severe shortages of heavy water during the war. In a practical application, the excess energy of the neutrons extracted by the moderator, and appearing as heat, in a reactor can be used to boil water to produce steam to drive an electric generating turbine in the production of electricity.

Bohr and Wheeler realized further that inevitably some of the reactor neutrons will be absorbed by U-238, taking them out of the reaction. But this process was still useful because, as before, the isotope U-239 decays, with a half-life of about 23 min, into element 93, neptunium. Researchers at the Berkeley Radiation Laboratory in

California first obtained this element in 1940. It too is unstable. It decays, they found, with a half-life of about 2.4 days into element 94, now called plutonium (after the then-planet beyond Neptune). It too was first isolated at Berkeley in 1940, but a paper reporting the discovery was withheld until after the war. Plutonium is long lived, relatively easily extracted by chemical means from a reactor, and it is even more fissionable than U-235. In essence, once a reactor is up and running it can be used to produce, not just heat to generate electricity, but plutonium to power a bomb. The world's first nuclear detonation, in the New Mexico desert on July 16, 1945, and the atomic bomb dropped over Nagasaki, Japan on August 9, 1945 were both powered by plutonium obtained from "breeder reactors" running for more than a year at the Hanford nuclear site in Washington state.

By the middle of 1939, much of the general theory regarding nuclear reactors and explosives had been published and became known to both sides of the coming war. But reactors and explosives still existed only in theory, and the theoretical predictions depended upon many measurements yet to be made regarding uranium and the fission process. In addition, overcoming the technological hurdles to obtaining the needed isotopes and constructing these devices remained a major difficulty then, as it does now.

The Early War Years

Physicists on both sides of the coming war immediately alerted their governments to the prospect of a nuclear weapon in 1939 (see Walker 1989). And on both sides they were greeted at first with skepticism. Well before American scientists convinced Albert Einstein, the nation's most famous scientist, to sign a letter to President Franklin D. Roosevelt in August 1939, warning him of the danger that Germany might obtain such a weapon, German researcher Paul Harteck and a colleague at the University of Hamburg had already informed Erich Schumann, the head of weapons research in the German Army Ordnance Bureau (*Heereswaffenamt*), of the prospect of a powerful new explosive. The skeptical Schumann (a descendent of the composer) handed the matter over to his explosives expert, Dr. Kurt Diebner, who was trained in physics. Diebner enlisted Dr. Erich Bagge, one of Heisenberg's former nuclear physics students, in a newly established military nuclear research effort. Thanks to Diebner and Bagge, by September 1, 1939, the day Hitler unleashed his panzer divisions into Poland, igniting the war in Europe, the world's first military nuclear fission project was already underway.

Within days of the outbreak of war, Diebner and Bagge issued military orders to Germany's leading nuclear scientists to attend meetings at the Army Ordnance Bureau in Berlin on September 16 and 26 to lay the groundwork for the war-time nuclear project. Among the attendees were Heisenberg, Hahn, Weizsäcker, Harteck, Hans Bothe, and Hans Geiger. They called themselves the Uranium Club. With army funding, the club members dispersed to their institutes to implement a multifaceted research program coordinated by Diebner and Bagge.

Heisenberg immediately threw himself into the work. Within three months he produced the first part of a two-part secret theoretical physics report to Army Ordnance on the practical applications of nuclear fission. Using the Bohr-Wheeler theory and the little available empirical data regarding uranium, he concentrated in his papers on the technical workings of a uranium "machine" (or "device"), a reactor that could drive German tanks and ships and power the German economy. In his first paper, he examined the usefulness of various moderators with different amounts of natural uranium in two geometrical reactor designs, spherical and cylindrical configurations of uranium and moderator arranged in alternating layers. He predicted that graphite and heavy water (water in which the hydrogen atoms possess an extra neutron in the nucleus) would prove the best materials to slow the neutrons without absorbing them so that they could go on to fission a rare U-235 nucleus rather than being captured by the more plentiful U-238. Heisenberg predicted that a controlled chain-reaction would require a layer configuration of 600 L of heavy water or 1000 kg of graphite, and 2000–3000 kg of uranium oxide. Heisenberg further predicted that by enriching the U-235 content of the uranium, a smaller reactor could be built to run at a higher temperature, suitable for powering war ships and large tanks. If one could separate enough nearly pure U-235 and compress it into a ball, the uncontrolled chain reaction would produce an incredibly powerful explosion. He urged the army to support isotope separation. Separation, he argued, was the "surest method" to obtain a working reactor, and, most importantly, it was "the only method for producing explosives" (Heisenberg 1942, 396). Heisenberg did not specify in this report how much U-235 was needed, the critical mass, in order to produce a nuclear explosion.

Heisenberg seemed less optimistic in the second part of his secret report about the practical realization of applied nuclear fission. The enrichment and separation of isotopes were beyond Germany's capabilities at that time. German industry still lacked techniques for processing the very dense uranium ore into the needed uranium plates and uranium oxide powder. Moreover, before the German capture of Europe's only heavy water production plant, in Norway, the German scientists did not yet have enough heavy water for reactor construction beyond small-scale testing. Heisenberg's mistaken conclusion that graphite would *not* be a suitable moderator for the reactor after all made the moderator problem much worse. Using imprecise data, Heisenberg calculated that graphite, a form of carbon, would absorb too many neutrons and thus halt the reaction. Further calculations by Weizsäcker's assistants at the Kaiser Wilhelm Institute for Physics supported this result. Even worse, actual measurements by Hans Bothe and co-workers in Heidelberg on neutron absorption in graphite seemed to confirm the conclusion against graphite. This situation resulted in the German scientists' fateful decision to ignore readily available graphite as a moderator and to rely instead solely on precious heavy water for building a reactor. Having captured and improved the Norwegian heavy water plant, Germany decided, in another fateful decision, not to build its own plant. The destruction of the Norwegian plant in 1943 in a daring Allied commando raid left

Germany without a steady source of heavy water, and its scarcity contributed to the failure of Germany's last attempts at the end of the war to achieve a reactor (Dahl 1999).

Unlike Allied researchers, the German researchers had failed to realize that even the purest industrial graphite still contained sufficient impurities to render it useless as a moderator. Only ultra-purified graphite would do. The world's first reactor, built by Enrico Fermi's team in Chicago, went critical in December 1942 with the use of uranium bricks interspersed with ultra-pure graphite bricks. If the Germans had utilized ultra-pure graphite and achieved an early reactor, one can only imagine what the German army might have done with the fissionable plutonium that the device would have produced.

Heisenberg's enthusiasm for quickly providing the army with two theoretical reports on the technical applications of nuclear fission derived from several concerns. Like most other German scientists and academics, he desired to defend and preserve German culture, which he and they regarded as independent of any regime that ruled Germany, and he likely saw this work as a contribution to that cause. He also wanted to control the project so as to keep decisions about it in his own hands rather than in those of Nazi ideologues dedicated to serving the German army. The backdrop of the previous seven-and-a-half years of the Third Reich offers further insight. Despite Himmler's "exoneration" of Heisenberg, Nazi officials still held him and other atomic scientists in suspicion. As noted earlier, by emphasizing the practical value of nuclear energy for the war effort, Heisenberg could demonstrate his own practical value and that of decent German physics, thereby rehabilitating himself and his science and silencing his ideological critics—all of which placed him in a tricky ethical and moral situation (for more, see Cassidy 2009, Chap. 22)

Through 1941 German fission research focused on confirming the details of Heisenberg's theoretical predictions for the two geometrical reactor designs and obtaining precise measurements of the properties and materials. Of the nine task-oriented research groups coordinated by Diebner and Bagge—in addition to one run by the German Post Office—Heisenberg worked closely with two groups focused on his experimental reactor designs. One was in his institute in Leipzig, the other was at the Kaiser Wilhelm Institute for Physics in Berlin. The Berlin institute had been built in 1936 with Rockefeller Foundation funds. The German army now seized control of the institute and replaced its Dutch director, Peter Debye, with the applied physicist Kurt Diebner. This only intensified the friction between Diebner and Heisenberg's closest Berlin colleagues, Weizsäcker and Karl Wirtz, who regarded Diebner as unworthy to administer their work. Through a bureaucratic maneuver they managed to replace Diebner with Heisenberg who was designated, in deference to Debye, director "at" but not "of" the institute. Diebner retreated to the army research station in Gottow, near Berlin, where he pursed his own reactor designs.

One of the German spherical tank reactors (AIP ESVA, Goudsmit Collection)

The Copenhagen Visit

Dividing his time equally between Leipzig and Berlin, Heisenberg oversaw a series of experiments in both places involving the two configurations of alternating layers of uranium metal powder and heavy water. Heisenberg and the Leipzig researchers tried concentric spherical shells of uranium powder in aluminum containers separated by heavy water. Because Heisenberg had garnered most of Germany's heavy-water supply for his Leipzig project, the Berlin team satisfied itself with ill-suited paraffin as a crude substitute moderator for a "pile" of circular, stacked containers of uranium powder between layers of paraffin in a cylindrical metal tank. The scientists then inserted a neutron source into the center of their contraptions and measured the number of neutrons emerging from the outside in the hope of finding an increase owing to fission. Not until the spring of 1942, after Heisenberg compensated for the loss of neutrons in his spherical aluminum containers, did the

Leipzig group detect the world's first neutron multiplication. Fermi soon caught up with and far surpassed that achievement.

German researchers also made progress on another front. In July 1940 Weizsäcker reported to the Ordnance Bureau that element 93, neptunium, produced when U-238 absorbs a neutron, should be as fissionable as U-235. The bombardment of individual nuclei of the plentiful isotope U-238 by two successive fast neutrons should result in a fission event: the first neutron produces U-239, which decays in 23 min into element 93; the second neutron then fissions the newly formed neptunium nucleus. A month earlier an American research team at Berkeley had already published in *Physical Review* the discovery that neptunium decays into the long-lived element 94, now called plutonium. Neptunium has a half-life of 2.4 days. In August 1941, German researcher Fritz Houtermans theoretically confirmed that plutonium is at least as fissionable as U-235 and that it could sustain a chain reaction yielding the explosive release of enormous amounts of energy. In addition, it should be more easily obtained than U-235 since it could be separated from the uranium in a working reactor by standard chemical methods.

Suddenly the road to an atomic bomb opened before the German researchers. It was no longer just a theoretical possibility. Heisenberg, Weizsäcker and their colleagues at the Kaiser Wilhelm Institute for Physics engaged in intensive discussions about this development and about whether or not they should proceed. Weizsäcker had recently returned from a visit to Copenhagen in German-occupied Denmark where he had met briefly with Bohr. All agreed that Heisenberg should talk with Bohr about the situation.

For those whose loyalty was suspect, permission to travel outside the Reich, or even to occupied countries within the extended Reich, was difficult to obtain. Through pressure exerted on party bureaucrats, and in part by Weizsäcker's father in the Reich's Foreign Office, Heisenberg and other prominent physicists were able to make a number of trips to German occupied countries, usually for the official purpose of cultural propaganda. But, to them, this was worth the opportunity to visit colleagues in occupied countries and worth the confirmation of personal and professional exoneration that such trips represented. Heisenberg's trips included visits to the United States in summer 1939, to occupied Poland and the Netherlands in 1943, and to occupied Denmark in 1944 after Bohr had fled the country. He traveled with Planck and Weizsäcker to Budapest in German-occupied Hungary in 1942, and with Weizsäcker to Copenhagen in September 1941 (for more, see Walker 1995). Supported by the senior Weizsäcker, the official propaganda purpose of the 1941 Copenhagen trip was to demonstrate the vibrancy of German science through a conference at the German Scientific Institute, a propaganda unit in Copenhagen that sought to attract Danish scientists. The unofficial purpose of the trip was for Heisenberg to discuss the problem of nuclear weapons with Niels Bohr.

Many historical analyses and Michael Frayn's popular theatrical play *Copenhagen* have explored Heisenberg's Copenhagen visit and the possible content of his unrecorded private conversations with Bohr (Frayn 1998). The visit has remained one of the most fervently debated issues surrounding Heisenberg and the German fission project. Whatever was actually said during Heisenberg's meetings with Bohr, the trip proved a disaster. Bohr received the impression that the Germans were working feverishly on a nuclear weapon, and he conveyed that impression to the Manhattan Project scientists. Heisenberg later argued that he had been completely misunderstood. In a postwar statement written in 1948, Heisenberg recalled opening their main conversation by asking whether Bohr believed "as a physicist one has the moral right to work on the practical exploitation of atomic energy." Shocked by the question, Bohr responded (in Heisenberg's statement) by asking whether Heisenberg believed that atomic energy could be exploited during this war. "Yes, I know that," Heisenberg replied. But he claimed that he was referring only to the reactor, not to a bomb. He further claimed that he told Bohr that the technical difficulties precluded construction of a bomb during the war.[6]

As much discussed in the literature, all of this is problematic, with many unanswered questions and alternative explanations. Was the morality of nuclear research really an issue for Heisenberg? He did not raise it in any surviving earlier documents. Bohr had often served in the past as a source of avuncular advice for the young Heisenberg, but in recent years Heisenberg consulted most closely with Planck and von Laue. Why didn't he approach them? What about two very different alternative explanations that arose after the war—that Heisenberg was attempting to stave off an Allied crash program to the build the bomb; and that Heisenberg was really on a spying mission to determine if the Allies were working on an atomic bomb, and, if so, how far along they were (e.g., Powers 2000; Rose 1998)?

In 1956, at the height of the Cold War and the nuclear arms race, Swiss journalist Robert Jungk published the German-language best seller *Heller als tausend Sonnen* (*Brighter Than a Thousand Suns*), his account of the history of the atomic bomb. Weizsäcker's assistance and perspective on the German effort were evident in Jungk's portrayal of the German researchers as having prevented the development of a German bomb out of moral scruples. Echoing one of Weizsackler's Farm Hall statements, Jungk wrote: "It seems paradoxical that the German physicists, living under a sabre-rattling dictatorship, obeyed the voice of conscience and attempted to prevent the construction of atom bombs, while their professional colleagues in the democracies, who had no coercion to fear, with very few exceptions concentrated their whole energies on production of the new weapon" (Jungk 1958, 105).

Acknowledging assistance from Weizsäcker, Jungk argued that the Germans deliberately maintained control of the fission project with the moral intent "to divert the minds of the National Socialist departments from the idea of so inhuman a weapon" (Jungk 1958, 88). Jungk then published in the Danish and English translations of his book an excerpt from a letter he received from Heisenberg giving Jungk his version of the Copenhagen visit in response to the German edition. Bohr attempted to respond to Heisenberg in a personal letter, which, however, after many drafts, was never sent. The Niels Bohr Archive released the drafts of this letter to

---

[6]Heisenberg, affidavit on the visit, draft and typescript, 1948. Submitted during Nuremberg trial of Ernst von Weizsäcker, in Heisenberg Papers, Archive of Max Planck Society, Berlin; and Weizsäcker defense exhibit 239, NARA, microfilm group M897, roll 119.

the public in 2002 in the wake of Frayn's play *Copenhagen*.[7] In these drafts Bohr contradicted and corrected nearly every point in Heisenberg's accounts of their meeting. In particular, he wrote, "If anything in my behavior could be interpreted as shock, it did not derive from such reports [of the possibility of a bomb] but rather from the news, as I had to understand it, that Germany was participating vigorously in a race to be the first with atomic weapons." It was that understanding that Bohr conveyed to the leaders of the Manhattan Project upon his escape from German-occupied Denmark in 1943. Heisenberg and Weizsäcker returned to Berlin to continue work on the German fission research project.

The Turning Point

Three months after Heisenberg and Weizsäcker returned to Berlin the German blitz across Europe had run its course. The invasion of Russia, begun in June 1941, bogged down outside Leningrad that winter. Hitler now ordered the complete mobilization of the German economy and all other resources for "total war." In response, Erich Schumann, head of army research, decreed that funding of any research would continue only "if a certainty exists of attaining an application in the foreseeable future."[8] An army evaluation of the uranium project reported in February 1942 that the reactor project might soon succeed but that a bomb depended on new isotope separation techniques or plutonium produced by a reactor.[9] Neither was assured in the foreseeable future.

The army immediately slashed funding for uranium research and relinquished control of the Kaiser Wilhelm Institute for Physics. It continued to provide a smaller budget in support of Diebner's rector research at the army's research station in Gottow, near Berlin. But the main project in Berlin was now suddenly up for grabs, and the scientists moved quickly to find acceptable support. Abraham Esau, head of the physics section in the Reich Research Council within the Reich Education Ministry, sought to gain control of the project and scheduled a series of non-technical lectures on nuclear development for Reich dignitaries on February 26, 1942. The speakers included the familiar figures: Heisenberg, Hahn, Bothe, Geiger, Harteck, as well as Klaus Clusius on isotope separation.[10] During the conference, chaired by Reich Education Minister Bernhard Rust, Heisenberg emphasized the positive prospects for reactor development and reviewed the possibility of a nuclear weapon, but he cautioned that the latter would not be ready in the foreseeable future owing to the difficulty of separating the rare fissionable

---

[7]The letters were published in Dörries (2005, 101–179).

[8]Directive of 5 Dec. 1941, quoted by Bagge, in Bagge et al. (1957, 28).

[9]"Energiegewinnung aus Uran," report to Army Ordnance, Feb. 1942 (Bagge Papers). I am grateful to Mark Walker for a copy of this report.

[10]Program in Goudsmit (1996, 169).

isotope U-235. However, he did tantalize his audience with the prospect that a working reactor would produce the easily separable element plutonium, a substance as fissionable as U-235. But it too would not be available in the foreseeable future.[11]

The lectures had the desired effect. The scientists could not be ordered to build a weapon that was beyond their immediate reach, and the Reich Education Ministry gained control of the project. It assigned the research to its subsidiary, the Reich Research Council, which placed it under Abraham Esau, the physics section head. Esau became the newly named Reich administrator (plenipotentiary) for nuclear physics. But that arrangement did not last long. The Kaiser Wilhelm Society, the network of government-sponsored research institutes, reclaimed its Berlin institute for physics, the main fission research center, and it managed to gain the support of Hitler's new armaments minister, Albert Speer, for this move. Speer induced Hitler to name Speer's boss, Luftwaffe commander Hermann Göring, as head of the Reich Research Council. At the same time, the German Physical Society used the contribution of nuclear physics to the war effort, along with Weizsäcker's diplomatic skills, to silence the remaining ideological opposition to theoretical physics and to Heisenberg in particular. In July 1942, Heisenberg was appointed professor of theoretical physics at the University of Berlin and official director of the Kaiser Wilhelm Institute for Physics, positions once occupied by Albert Einstein. The prestigious dual appointment fulfilled Himmler's long-standing promise of proof of "exoneration." A year later, Heisenberg persuaded Speer and Göring to replace Esau with his long-time Munich colleague Walther Gerlach as Reich administrator of fission research.

Later at Farm Hall, Heisenberg recalled the success of the February 1942 lecture series in glowing terms: "One can say that the first time large funds were made available in Germany was in the spring of 1942 after that meeting with Rust when we convinced him that we had absolutely definite proof that it could be done" (Farm Hall 1993, 76). "It" may have referred to the bomb, in the long term, but it probably referred to the reactor, in the short term.

Still, the new funds were not that large, and Heisenberg had to give further presentations to Speer and military commanders in June 1942, and to Göring and assembled dignitaries at the Aeronautical Research Academy in 1943, in order to maintain their support. In both instances, he carefully laid out the possibilities of a nuclear reactor and an explosive, along with his optimism for the former and his pessimism for the latter in the foreseeable future. As discussed later at Farm Hall, he and colleagues did not want to be ordered to build the bomb; since failure to do so at the height of war would surely have meant execution. During his talk before Göring's academy in 1943 Heisenberg left the plutonium alternative unmentioned.

---

[11]Heisenberg, lecture on 26 Feb. 1945, published in Heisenberg (1989a, 517–521).

The Later War Years

As the new scientific head of Germany's main fission research project, starting in July 1942, Heisenberg laid plans for the large-scale construction of a fully working reactor at the Berlin institute based on the results of the preliminary small-scale experiments in Leipzig and Berlin. Going far beyond his predictions in 1940, he submitted an order in July 1942 to the German industrial giant the Auer Company for three metric tons (3.3 English tons) of rolled uranium plates. He envisioned a reactor design consisting of the metal plates alternating with 1.5 tons of heavy water in a cylindrical tank. However, in the same month, Diebner at Gottow submitted his own order for uranium metal *cubes* to be used in an alternate design in which the cubes were suspended on aluminum chains in a tank of heavy water. The greater surface area of uranium exposed to the heavy water would enable the fission of more atoms. But Heisenberg's order took precedence.

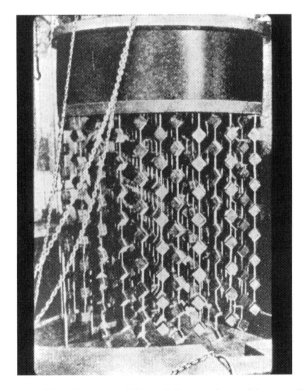

Uranium cube reactor. The cubes suspended on chains were lowered into a cylindrical tank of heavy water as a moderator (AIP ESVA, Goudsmit Collection)

As the Allied bombing campaign mounted over Germany, Heisenberg's uranium did not arrive until nearly a year later. Now fully exonerated and the fission project safely underway as a demonstration to friend and foe alike of the utility of nuclear research and the continued viability of German physics, Heisenberg used the hiatus in his nuclear engineering work to pursue his primary interest, theoretical quantum physics. He began submitting a series of theoretical physics papers for publication in a German journal on the scattering of high-energy elementary particles. These led to his well-known S-matrix theory of particle scattering, which enjoyed considerable international interest after the war.

During this same period Bagge in Berlin and Clusius in Munich pursued various methods for separating the fissionable isotope U-235, but with little success. Not so Diebner. After his uranium cubes finally arrived, Diebner made remarkable progress with his design, obtaining neutron multiplication far greater than Heisenberg's Leipzig group had ever achieved. By that time the army had withdrawn entirely from nuclear research, and his project was under the auspices of the Reich Bureau of Standards. When it soon came under the oversight of nuclear administrator Walther Gerlach, Gerlach began shifting Research Council funds to Diebner's reactor, along with additional funds he managed to obtain from the SS. He allowed Diebner's and Heisenberg's projects to proceed simultaneously in the hope that at least one would succeed. He apparently overlooked the likelihood that the competition for scarce resources would prevent either from achieving success.

By the time the uranium plates arrived in Berlin, Heisenberg was ready to concede the superiority of Diebner's design. But the Allied bombing raids had become so intense that during the summer of 1943 Speer ordered all research groups to begin vacating the city. As Heisenberg sent the plates back to Auer to be cut into cubes, nearly a third of his institute of 55 members fled Berlin for a vacant textile factory in the Swabian village of Hechingen nestled among the rolling hills of southwestern Germany near the Black Forest. Members of Otto Hahn's institute for chemistry, now in ruins from the bombing, moved to the nearby town of Tailfingen. They were soon joined by Carl Friedrich von Weizsäcker who had escaped capture and the bombing of German-occupied Strasbourg in France, where he had accepted a professorship in theoretical physics. As the Allied bombing of German cities intensified, Heisenberg moved his wife and growing family of six children permanently to their summer cottage in Urfeld near the shore of Lake Walchen in Upper Bavaria, south of Munich.

Heisenberg and Wirtz bravely remained at their posts in Berlin through the end of 1944. They were in the midst of assembling their uranium and heavy water into a long-awaited reactor when Gerlach suddenly ordered them out of the city. The city was nearly under siege, with massive air raids from above and the Soviet Red Army rolling relentlessly westward toward Berlin. Gerlach arranged for Diebner and his equipment to join Heisenberg's convoy of trucks south, but half way to Hechingen Gerlach had second thoughts. When the convoy arrived in the town of Stadtilm in the Thüringen Forest, about half way to Hechingen, Gerlach insisted on unloading Diebner's equipment and material and setting up Diebner's project in a last-ditch attempt to achieve the reactor. After much confusion, Heisenberg and Wirtz finally arrived in Hechingen, where all available hands worked feverishly in the nearby town of Haigerloch within a small cave in the side of a huge rock, safe from bombing, in one last attempt to achieve a self-sustaining nuclear reactor. Neither group had sufficient heavy water for the task. Together they might well have succeeded had Gerlach not made the fateful decision to stop in Stadtilm. For Heisenberg, success would have demonstrated the survival of decent German physics and, perhaps equally important, would have made German physicists influential figures in the postwar reconstruction of Germany. They were, after all, still far ahead of Allied fission research—weren't they?

The former wine cellar built into the side of a huge rock beneath a church in Haigerloch where Heisenberg's team assembled their last reactor (Photograph by Samuel Goudsmit, courtesy of AIP ESVA, Goudsmit Collection)

Just as the last reactor attempts neared completion at the end of April and early May 1945, the American led Alsos Mission swept across Germany and snatched up all of the scientists, together with their papers and equipment. The German nuclear fission project was over. A devastated Germany surrendered unconditionally on May 8, 1945.

The Alsos Mission

The covert alliances, scientific secrets, and rapid technological advances characterizing the Second World War brought an increasing need for "field operations" to gain more "intelligence" about what others were up to. In the United States, numerous military and civilian agencies at various levels, such as the Office for Strategic Services (OSS), the precursor to the CIA, arose to meet the challenge. Many such agencies included nuclear intelligence among their activities. In order to untangle the oversight responsibility for nuclear affairs, in the fall of 1943 US Army Chief of Staff Gen. George C. Marshall assigned all nuclear intelligence to the Manhattan Project, which was under the military command of Maj. General Leslie R. Groves. Following the successful Allied invasion of Italy that September, Groves dispatched into Italy, from North Africa, a secret nuclear intelligence operation called the Alsos Mission—named after the Greek word "alsos" for grove —under the command of Army Colonel Boris T. Pash, a Russian-American known for his bravado. Enrico Fermi's institute in Rome had been a center of pre-war nuclear research before Fermi fled to the United States. Many of his co-workers had remained, and some were in contact with German physicists.

Pash's Alsos Mission came up empty handed. The problem was the lack of scientific advice to guide their activities. As the United States prepared for the D-Day invasion of northern Europe in 1944, Groves prepared to dispatch a second Alsos Mission into Europe from England. It's mission: to track down and disable a possible German atomic bomb and to capture the German nuclear scientists and their equipment. Army G-2 (intelligence) and Naval Intelligence, the only US military services at the time, again selected Pash and assigned other military personnel for the new mission. This time the mission also included a staff of scientific advisors to guide the operation. Groves turned to Vannevar Bush, the director of the Office for Scientific Research and Development (OSRD) which coordinated all civilian research for the war effort, for the recommendation of a chief scientist to lead the civilian "expert consultants." (Bush later founded the National Science Foundation.) On May 2, 1944, the Assistant Chief of Staff of Army G-2, General Clayton Bissell, who coordinated with OSRD, wrote to inform Groves that OSRD had only a single recommendation: Dr. Samuel A. Goudsmit.[12]

---

[12]Bissell to Groves, memo, 2 May 1944, NARA, RG 77 (Manhattan Engineer District), Microfilm Group M1109, roll 4 (Alsos Mission).

Samuel A. Goudsmit at his desk during the Alsos period (AIP ESVA, Gift of Michaele Thurgood Haynes and Terry Thurgood, Thurgood Collection)

Goudsmit had received his doctorate in physics with Paul Ehrenfest in Leiden. He was best known for introducing, together with his Dutch countryman George Uhlenbeck, the essential notion of electron spin shortly after Heisenberg had made the breakthrough to quantum mechanics in 1925. He moved to the United States in 1927 as a physics professor at the University of Michigan and obtained United States citizenship. While at Michigan he met with his European colleagues, including Heisenberg, during the annual Michigan international summer schools on theoretical physics. The last of these occurred in summer 1939, just months after the discovery of fission and just weeks before the outbreak of war in Europe. During that meeting Goudsmit, Max Dresden and other physicists attempted in vain to persuade Heisenberg to remain in the United States during the coming war. Heisenberg refused with the simple reply, "Germany needs me" (Goudsmit 1996, 112).

By 1944 Goudsmit was already accustomed to transmitting to US authorities any intelligence he gained through his European connections. In November 1942, he informed an official that Heisenberg had become director of the Berlin physics institute and that he was working on "tube alloys," the British code phrase for the atomic bomb.[13] Before Groves received his nuclear intelligence mandate, Goudsmit wrote to Caltech physicist Lee DuBridge, head of radar research, to volunteer his services for scientific intelligence.

> No doubt Europe will crack up pretty soon, probably in parts. It would be advantageous to us to get information on the work of the scientists over there as soon as possible. Now, I have very close personal contacts with most of the physicists in Italy, France, Belgium, Holland and even Germany. I think there are even some German physicists who still believe I am their friend. I might be mistaken, of course. Anyway, I have a feeling that my very close personal connections may be helpful in obtaining information which would not be revealed to a committee of men they are not so well acquainted with.[14]

When the OSRD submitted its sole recommendation of Goudsmit he had been working on radar at the MIT Radiation Laboratory since its inception. His familiarity with the Manhattan Project was only from a distance. That would be an advantage should he be captured and interrogated by the enemy. In addition, he was multi-lingual. Goudsmit seemed the obvious choice for chief Alsos scientist, but General Bissell added rather derisively in his letter to Groves, "Dr. Goudsmit has been recommended principally because he is available and is not presently assigned to important work. Dr. Bush has no other names to suggest. It is my opinion that his name adds nothing to the prestige of the mission."[15] The "can do" Gen. Groves valued success over prestige and assigned Goudsmit to the task. The OSRD placed him and other staff scientists under the control of its Office of Field Service, directed by MIT president Karl T. Compton, the brother of Nobel Prize physicist Arthur H. Compton. On August 18, 1944 two months after the D-Day invasion of Europe, Compton ordered Goudsmit, then in Washington, to report immediately to the London office of OSRD for "assignment to the Assistant Chief of Staff, G-2, ETOUSA [European Theatre of Operations, US Army] for duties with the Scientific Intelligence (Alsos) Mission."[16] Six days later he and two other civilian scientists (William Noyes and Clarence Hickman), now outfitted in London with military uniforms and armed with cameras to shoot pictures, received military orders to proceed from London to "points within the combat zone in Northwestern Europe, on duty as indicated, as may be necessary to carry out the orders and the verbal instructions of the commanding general."[17]

---

[13]Goudsmit to W.B. Lewis, 7 Nov. 1942, Goudsmit Papers, AIP NBLA, box 25, folder 3. The online files are organized according to the locations of the physical documents.

[14]Goudsmit to Lee DuBridge, 25 June 1943, ibid.

[15]Note 12.

[16]K.T. Compton to Goudsmit, 18 Aug. 1944, Goudsmit Papers, AIP NBLA, box 25, folder 3.

[17]Capt. Gallup to Goudsmit, 24 Aug. 1944, ibid.

Col. Pash giving orders to advance through Thanheim on the way to southern Germany (Michaele Thurgood Haynes and Terry Thurgood, courtesy of NBLA)

Goudsmit standing in a M-20 armored car at Stadtilm, Germany (AIP ESVA, Goudsmit Collection)

As the combat zone moved eastward through northwestern Europe, the Alsos Mission followed close behind under the military command of Col. Pash and the scientific command of "expert consultant" Dr. Goudsmit. Goudsmit later provided an insightful account of their work in his popular book *Alsos*, as did Col. Pash in his less well-known account, *The Alsos Mission* (Pash 1969, see also Mahoney 1981). After American and British forces swept through Belgium, Goudsmit and his staff of scientists eagerly examined documents in the local offices of the Auer Company. The German company had obtained and processed the uranium from Belgium's African colonies used by the German project. At about that time, Goudsmit took a brief trip from Brussels to his boyhood home in The Hague where his parents had been arrested and sent to their deaths at Auschwitz. Standing alone in the ruins of his boyhood home, Goudsmit was overcome with grief, anger at the Germans, and guilt that he had not removed them from danger in time (Goudsmit 1996, 46–48). It was a story that must have been repeated many times by others (and it inspired Scene 15 of *Farm Hall*).

On November 15, Allied forces liberated Strasbourg, France where Weizsäcker held forth as physics professor. The professor had fled, but he left behind some of his valuable uranium project papers. From these Goudsmit determined that Heisenberg and his project had evacuated Berlin for the town of Hechingen in southern Germany. From a piece of project stationery, Goudsmit obtained Heisenberg's address and even his telephone number. He badly wanted to pick up the telephone and give Heisenberg a call, but the lines were down, missing what would have surely been a stunning high point to the story. Most importantly, from the captured papers Goudsmit was able to report to Groves his preliminary conclusion that the Germans did not have a nuclear explosive. While awaiting more evidence, Groves spurred the Manhattan Project on in its urgent dash to build the bomb.

Lt. Col. John Lansdale (*left*) and Col. Pash at the Institute for Physics, Strasbourg, where Weizsäcker had worked (Michaele Thurgood Haynes and Terry Thurgood, courtesy of NBLA)

Lt. Toepel (*left*) and Samuel A. Goudsmit (*right*) in Stadtilm (Photograph by Malcolm Thurgood, courtesy of AIP ESVA)

Alsos Mission members examining papers in Gerlach's office in Stadtilm. From *left* to *right* Samuel A. Goudsmit, Fred A. Wardenberg, and British intelligence agents Lt. Cmdr. Eric Welch and Wing Cmdr. Rupert A. Cecil (Brookhaven National Laboratory, courtesy of AIP ESVA)

Following the Battle of the Bulge in December 1944, American and British forces began fighting their way to the Rhine and crossed into Germany in March 1945. Again the Alsos Mission followed close behind. Although the mission fell under the military command of ETOUSA, G-2, it operated under the intelligence guidance of the recently formed Combined [American and British] Intelligence Objectives Subcommittee (CIOS). This brought an expansion of the mission staff to include British scientists, intelligence agents, and soldiers (Mahoney 1981). According to Alsos personnel lists in the Goudsmit papers and in the US National Archives, from early 1945 through the armistice in May, the Alsos Mission was composed of 51 civilian "expert consultants," four other civilian personnel, six CIC (Counter Intelligence Corps) agents, and 119 military personnel. Besides Goudsmit, the American expert science consultants included Harvard physicist Edwin C. Kemble (deputy chief) and Michigan physicist Walter F. Colby (assistant chief). Pash's military staff included Lt. Col. George R. Eckham (deputy chief), Maj. Richard C. Ham (operations) and Maj. Robert Furman. Among the British contingent were Sir Michael Perrin (scientist) and the military commanders Lt. Col. Sir Charles Hambro and Lt. Commander Eric Welsh (Navy).[18] Many of these men played important roles in the Farm Hall story.

Among the Alsos Mission's military staff was the mission's official photographer, Army Sgt. Malcolm (Mickey) Thurgood from Indiana. Surprisingly, he was accepted into the Army in 1942 despite his flat feet and near blindness in one eye. Moreover, after he completed basic training he was sent for specialized photography training, then transferred to London to join the Alsos Mission.[19] A selection from his many photographs appears in this book.

---

[18]Goudsmit Papers, AIP NBLA, box 25, folder 2.

[19]I am grateful to Michaele Thurgood Haynes and Terry Thurgood for this information.

Sgt. Malcolm Thurgood, Alsos Mission photographer, with his favorite 4 × 5 camera (Michaele Thurgood Haynes and Terry Thurgood)

Sgt. Malcolm Thurgood, May 1945 (AIP ESVA, Gift of Michaele Thurgood Haynes and Terry Thurgood, Thurgood Collection)

As the Alsos Mission moved into Germany Col. Pash established two western staging areas: Advance Base North in Aachen, in the future British zone of occupation, which targeted Diebner's reactor project in Stadtilm, among others; and, on March 30, 1945, Advance Base South in Heidelberg, in the future American zone of occupation. From Heidelberg, Pash and Goudsmit prepared to descend on the Swabian towns of Hechingen and Haigerloch where Heisenberg's institute and reactor project was located, and Tailfingen where Hahn's chemistry institute had settled. On April 8, 1945, US Secretary of War Henry L. Stimson, Army Chief of Staff Gen. Marshall, and Manhattan Project commander Gen. Groves agreed that the German scientists and their equipment should be captured immediately. But the southern region was still in enemy hands. According to a report to Gen. Groves from Alsos military member Lt. Col. John Lansdale, Pash had requested an

American General Bull to undertake an immediate invasion of the area. Bull responded that it would require at least three army divisions. "He stated, however," wrote Lansdale, "that he could bomb the area, and, also, that when word was received that the French were moving in, an Airborne division could be sent in, in support of the French. I requested that he contemplate the use of both methods and that our feeling was that the individuals and materials down there should be seized by the Americans in advance of the French, or if that were impossible, destroyed to fullest extent."[20] The Americans did not wish the French or the Russians to gain control of these scientists or their nuclear project.

The French advanced faster than expected and continued right through their assigned target area into the future American zone. Pash immediately jumped into action. Lansdale offered a vivid description of the operation that began shortly after the French advance on April 23. Gen. Devens immediately assigned to Col. Pash a combat engineer battalion which moved out at once for the target area. This force arrived at Haigerloch within an hour after the arrival of the French. At Haigerloch, they secured the targets and put them under guard, then moved on to Hechingen, where the same thing was done, also within about an hour after the arrival of the French.

---

[20]Lansdale to Groves, 5 May 1945, NARA, RG 77, Microfilm M1109, Roll 2 (Alsos Mission).

Lt. Fiebig (*left*), operations officer, confers with Col. Pash (*right*) at Haigerloch (Michaele Thurgood Haynes and Terry Thurgood, courtesy of NBLA)

The capture of Otto Hahn outside his laboratory in Tailfingen (Michaele Thurgood Haynes and Terry Thurgood, NBLA)

Lansdale was still in London at that time. Accompanied by British agents, he immediately flew to Heidelberg, where the US Army also had a military air base, then he headed to Haigerloch by jeep, the location of Heisenberg's reactor. Lansdale continued:

On 24 April we explored the laboratory at Haigerloch and started dismantling the pile. I proceeded on to Hechingen that evening, leaving the British party in Haigerloch. That evening, we received word that the French were to move into Tailfingen. On 25 April, we went with party to Tailfingen, accompanied by General Harrison, G-2, 6th Army Group. The party arrived in Tailfingen in advance of the French and occupied the town. All targets were secured and Otto Hahn turned over all papers to us.

In the meantime, the British party, with the assistance of Lt. Parrish, and a platoon of engineers, dismantled and loaded the pile on trucks. The outer shell of the pile was destroyed by hand grenades.

On 26 April, with Welsh and Perrin, interrogated Von Laue, Von Weizsäcker, Wirtz and Hahn. After much questioning, Wirtz agreed to show us the place where the heavy water and material from the Haigerloch pile had been taken. We obtained the material and loaded it on a truck, dispatching it to Paris. On 27 April, sent the prisoners off to Heidelberg. Just before departure, Weizsäcker told us the hiding places of his papers, which were obtained. On 28 April, went to Reims [SHAEF headquarters] with the British party. Here Sir Charles Hambro, Furman and I called on General Strong and arranged for the housing and care of the prisoners. We also confirmed the arrangements for the transportation of the material from Hildesheim to England, on which Sir Charles Hambro and Furman had been working in the meantime.[21]

_____

[21]Ibid.

Alsos members dismantling the German reactor at Haigerloch. Standing, *left* to *right* Lt. Col. Lansdale and Lt. Cmdr. Welch. Foreground: Wing Cmdr. Cecil, and behind him Sir Michael Perrin (Brookhaven National Laboratory, courtesy of AIP ESVA)

Alsos members looking for buried uranium cubes near Haigerloch. ESVA identification: from *left* to *right* Sir Michael Perrin, Samuel A. Goudsmit, Lt. Col. John Lansdale, Lt Cmdr. Eric Welch (Brookhaven National Laboratory, courtesy of AIP ESVA)

Alsos members digging up a hidden chest of uranium cubes near Haigerloch (Photograph by Samuel Goudsmit, courtesy of AIP ESVA, Goudsmit Collection)

Lansdale did not mention that Goudsmit and the Alsos Mission followed close behind and helped dismantle the Haigerloch pile on April 24. However, Goudsmit reported, he was not present when Pash and Gen. Harrison, the regional intelligence chief, entered Hechingen and quickly proceeded into Heisenberg's office. Heisenberg was not there, but they did find on his desk a prominently displayed photograph of Heisenberg and Goudsmit standing together side by side. The photograph had been taken in 1939 when Heisenberg stayed with Goudsmit in Ann Arbor, Michigan during his last visit to the United States. It is not known if Heisenberg purposely left it there, but the impish colonel nearly convinced the intelligence general that Goudsmit could not be trusted because of his close contact with the enemy. "I had to stand a lot of teasing about the incident afterwards," Goudsmit wrote (Goudsmit 1996, 99) (The incident inspired the photograph episode in Scene 15 of *Farm Hall*).

Alsos members meet with Col. Pash in Heisenberg's institute in Hechingen. Among them are: Back row: Lt. Cmdr. Welch (2nd from *left*). Front row, right to *left* Col. Pash, James A. Lane, Sir Michael Perrin, Lt. Col. Lansdale, Fred A. Wardenberg (Brookhaven National Laboratory, courtesy of AIP ESVA)

Also absent from Lansdale's account of the operation is any mention of the Alsos Mission's top three "targets," Diebner, Gerlach, and "target number 1": Heisenberg. None was in the area at that time. Diebner had fled from his reactor project on April 8 as Gen. Patton's tanks, rolling rapidly across Germany, closed in on Stadtilm. He escaped south to join Gerlach, who had already abandoned the outpost for his home and laboratory in Munich. Heisenberg had a more harrowing experience. As news of the French advance reached Haigerloch, Heisenberg ensured that the project's equipment and papers were safely hidden for later recovery, then he set out on a wild bicycle ride across war-torn southern Germany in an attempt to reach his family in their mountain cabin in southern Bavaria. Dodging Allied aircraft strafing and marauding bands of army and SS soldiers, Heisenberg somehow made it to the cabin, where he awaited capture while preparing his family for his inevitable absence.[22]

---

[22]Heisenberg (1971, 190–191), and Heisenberg's dairy, 15 Apr. to 3 May 1945, in Heisenberg and Heisenberg (2016, 250–263).

Learning through interrogations about the locations of the three missing targets, on April 20 the Heidelberg base (now the mission's main base) dispatched two "operational teams" to capture the targets. The first, under the command of Maj. Ham, had little difficulty locating Gerlach in Munich. He had quietly resumed his pre-war work in his university laboratory. They took him to his home for interrogation, which led them to Diebner in a small town south of Munich.[23] Both targets and their papers were dispatched to Heidelberg on May 3.

The other team, commanded by Col. Pash, faced greater difficulties reaching target number 1. Heisenberg's location in Urfeld was still behind enemy lines and nestled in difficult mountainous terrain. Pash offered a thrilling account of his "alpine operation" in a report to the War Department in Washington and in his book *The Alsos Mission*.[24] Despite the enemy units in the area, on May 3 Pash, brandishing a pistol in command of a small unit of soldiers, arrived at the Heisenberg cabin to find their target on the front porch, blissfully awaiting their arrival. As Heisenberg's wife, mother, a neighbor woman, the woman's child, and Heisenberg's six wide-eyed children looked on, they arrested their target, searched his house, and seized his papers. Among the papers was a batch of especially significant scientific letters to Heisenberg from leading physicists that he had wanted to keep safe.[25] The letters have not been seen since.

At 8 PM, the Alpine Operation placed their captive in the back of Pash's jeep, and the tiny force set out for a night-long journey back to Heidelberg. Bouncing over heavily bombed out roads, they finally arrived in Heidelberg on the morning of May 4 with everyone complaining of sore backs. Three days later the surviving commanders of the Third Reich surrendered at Allied headquarters in Reims, France, officially ending the war, the Reich, and the German uranium project.

The Meager Progress

Following the discovery of nuclear fission in Berlin and publication of the discovery in early 1939, Germany was the first nation to establish a military project for the exploitation of nuclear fission. In July 1939, its scientists published a paper indicating that research had already begun. Soon after the outbreak of war in September 1939, the German army captured uranium supplies in Belgium and uranium mines in Czechoslovakia. The army also gained control of the Norwegian heavy water plant, Europe's only such plant, and Germany's Uranium Club was already underway before the end of September. No wonder American scientists, turning to Einstein, the most prominent physicist in the United States, found it

---

[23]Major Ham to Colonel Pash, Report, 12 May 45, NARA, RG 77, Microfilm M1109, roll 4.

[24]Pash to Chief, Military Intelligence Service, War Dept., "Subject: Alpine Operation," 18 May 45, NARA, RG 77, M1109, roll 4; and Pash (1969, 219–241).

[25]Interview with Mrs. Heisenberg, Göttingen, Germany.

necessary to warn the president of the danger that Germany was far ahead in the construction of a possible nuclear weapon.

Five years later, when the Alsos Mission put an end to the German uranium project, the Germans had not even achieved a controlled chain reaction, a working reactor. Enrico Fermi had done so in Chicago nearly three years earlier. What happened? Why were the results of the German project, which had been so far ahead, so meager by the end of the war in comparison with the achievement of the Manhattan Project? Answers to these questions have been the subject of considerable, sometimes heated, debate ever since the end of the war. The answers, often argued with unwavering insistence, have ranged across the entire spectrum of possibilities: from the scientists' incompetence, to the hindrances caused by German anti-Semitic policies and the war-time bombing, to the impossibility that war-time Germany could have launched a project on the scale of the Manhattan Project, to the scientists' sabotage of the project because of moral scruples about giving a nuclear weapon to Hitler. Others have argued that the war-time conditions and the German army's decision to drop the project in 1942 precluded anything more than meager progress, while the argument has been made that Heisenberg's decision to focus instead on his quantum theoretical research over nuclear engineering, once he had achieved his policy goals for the nuclear project in 1942, further hampered progress thereafter (see Bibliography).

Samuel A. Goudsmit presented the first comprehensive assessment of the German project and its leaders in his 1947 book *Alsos* (reprint Goudsmit 1996). Their failure, he argued, arose from a combination of individual failings and other factors: the Nazi regime's destruction of the once mighty German physics through anti-Semitism and ideology, the confused organization of German science under the Third Reich, the scientists' complacency owing to a false belief in their continued scientific and technological superiority over the Allies, and Heisenberg's scientific errors regarding graphite, a chain reaction and the critical mass. Even worse, wrote Goudsmit, the scientists engaged in an excess of hero worship, such as that practiced by "the smug Heisenberg clique," that overlooked more practical-minded technicians such as Kurt Diebner (Goudsmit 1996, 121).

Heisenberg vehemently challenged Goudsmit's assessment in long letters exchanged with Goudsmit and in a series of letters and interviews published in *The New York Times*. Their extensive correspondence is now among the most popular documents in the Goudsmit Collection.[26] For Heisenberg and his colleagues, their reputations as well as their roles in the revival of West German science were at stake. The two men never resolved their differences. Heisenberg and the other leading figures at Farm Hall went on to highly successful carriers in science and science policy in the newly emerging West German Federal Republic (McPartland 2013).

Goudsmit's assessment of the German scientists, and their project, found its counterpoint a decade later in Robert Jungk's popular book *Brighter Than a Thousand Suns*. These two books have remained influential ever since. As we have

---

[26]Samuel A. Goudsmit Papers, AIP NBLA, Box 10.

seen, Carl Friedrich von Weizsäcker encouraged Jungk's assertions regarding morality and deliberate behavior, which Weizsäcker had earlier expressed during urgent discussions among the scientists at Farm Hall shortly after learning of the first atomic bomb dropped on Japan. These discussions culminated on August 7 in the Farm Hall Memorandum (depicted in the play *Farm Hall*). On that day Max von Laue wrote a letter to his son in Princeton, which was sent after the authorities granted permission to send letters. In it Laue summarized the position that the scientists had formulated at Farm Hall for the memorandum. After outlining the practical difficulties that hindered such work, he then alluded to what became the moral argument:

> Our entire uranium research was directed toward achieving a uranium machine as an energy source, first because no one believed in the possibility of a bomb in the foreseeable future; second, because basically no one among us wanted to put such a weapon in Hitler's hands.[27]

Fourteen years later, on April 4, 1959, Laue wrote a letter to his colleague Paul Rosbaud in which, after referring to Jungk's book, he recalled the discussion that day and two weeks later when Heisenberg gave a lecture in which he finally made a correct calculation of the critical mass.

> After that day we talked much about the conditions of an atomic explosion. Heisenberg gave a lecture on the subject in one of the colloquia which we prisoners had arranged for ourselves. Later, during the table conversation, the version (*Lesart*) was developed that the German atomic physicists really had not wanted the atomic bomb, either because it was impossible to achieve it during the expected duration of the war or because they simply did not want to have it at all. The leader in these discussions was Weizsäcker. I did not hear the mention of any ethical point of view. Heisenberg was mostly silent (Bernstein 2001, 352–353).

Although Heisenberg did state at Farm Hall, "we were not 100% anxious to do it," he did not say they had refused to do it. Weizsäcker admitted much later that they were indeed working toward an explosive device during the early years, before the army dropped the project in early 1942.[28]

---

[27]Max von Laue to Theodor von Laue, 7 Aug. 1945, Laue Papers.

[28]C.F. von Weizsäcker, *Bewusstseinswandel* (Munich: Carl Hanser Verlag, 1988, 365), excerpt from an interview with *Stern*.

Building in Heidelberg where the captive German scientists were held before transfer to a prisoner
of war facility (AIP ESVA)

Goudsmit (*right*) working with Alsos staff members (AIP ESVA, Gift of Michaele Thurgood Haynes and Terry Thurgood, Thurgood Collection)

Farm Hall

Samuel A. Goudsmit, along with his staff of scientists and intelligence officers, interrogated the atomic scientists as they were captured and brought to the Heidelberg headquarters. They also thoroughly studied their papers. The equipment and papers went to intelligence authorities in Paris, while the scientists were whisked into a detention center called "Dustbin" at the Chateau du Chesnay near Versailles. On May 1, 1945, Lt. Cdr. Welsh ordered British Army intelligence agent Maj. T. H. Rittner to travel from London to Gen. Eisenhower's SHAEF head-quarters in Reims (or Rheims), France in order to take charge of a "party" of captive German scientists.[29] This was shortly before the capture of Heisenberg, Gerlach, Harteck, and Diebner. In a retrospective "Preamble" to his subsequent secret Farm Hall Reports to Gen. Groves and his British superior officers, Rittner began with the following (unedited) account (Farm Hall 1993, 19):

---

[29]Little is known of T. H. Rittner, even his first and middle names are not known for certain.

1st May 1945

I received at H.Q. from Lieut. Cdr. WELSH instructions to proceed to RHEIMS (France). to report to G2 SHAEF and collect a party of German Scientists. A Chateau at SPA (Belgium) had been prepared for their detention. A number of distinguished British and American Scientists would be visiting them in the near future and my instructions were that these Germans were to be treated as guests. No one, repeat no one, was to contact them except on instructions from H.Q.

2nd May 1945

I proceeded by air to RHEIMS and reported to SHAEF where I was informed by Major KEITH, P.A. to A.C. of S. [Personal Assistant to Assistant Army Chief of Staff] G2 that the Chateau at SPA was no longer available and that the party was to be held at RHEIMS at 75 Rue Gambetta until other arrangements could be made.

Arrangements had been made to draw American "A" Rations ready cooked and a staff of two British Orderlies and an American cook had been provided by SHAEF in addition to the necessary guards.

The same evening. The following arrived at 75, Rue Gambetta, escorted by Major FURMAN, U.S. Army:

> Professor HAHN
> Professor VON LAUE
> Doctor   VON WEIZSACKER
> Doctor   WIRTZ
> Doctor   BAGGE
> Doctor   KORSCHING.

Professor MATTAUCH, whom I had been told to expect was not among the party. The professors were friendly and settled down well. They expressed appreciation of the good treatment they were receiving and a very pleasant atmosphere prevailed. At my request they gave me their personal parole not to leave the house or that portion of the garden which I allotted to them.

9th May 1945

Major FURMAN informed me that he was arranging for us to return to RHEIMS and that in the meantime the following Germans were to join the party:

> Professor HEISENBERG
> Doctor DIEBNER.

Rittner did not mention when Gerlach and Harteck arrived. His description raises a number of questions: Why was a British Army major put in charge of these high-profile captives held by the Americans? Why were they treated as "guests"? Why were they held incommunicado? Why was their word of honor considered sufficient to prevent their escape? Why were these ten selected? Others, such as Josef Mattauch, Klaus Clusius and Hans Bothe had contributed significantly to the fission research, while Otto Hahn and Max von Laue had much less, if anything, to do with the war-time project.

For German professors, who held integrity among the highest virtues in research and life, one's word of honor could not be broken. Upon arrival at Farm Hall, Rittner required his "guests" to repeat their word of honor in writing. Several

background materials help provide answers to the other questions. British physicist and science intelligence expert Reginald Victor Jones (who preferred R.V. Jones) played an influential advisory role to the Alsos Mission. According to his memoir, he took a special interest in the Alsos captives after Lt. Cdr. Welsh informed him that an American general had expressed the opinion that the best solution was to shoot all of the German nuclear physicists (Jones 1978, 481). It is not known which general said this or when, but it prompted Jones, Welsh and Perrin to arrange for British authorities to take control of the scientists from the Americans as soon as possible—and for Maj. Rittner, further down the chain of command, to assume direct responsibility for them.

Goudsmit explained in *Alsos* that he and the Americans had selected these ten captives because they included what they regarded as the eight leading figures in the German fission project. They also included Hahn and Laue in anticipation of their future contribution to the post-war revival of German science (Goudsmit 1996, 103–106). During those early postwar months, the Americans and British were concerned to keep the scientists from contacting other Germans who might engage in sabotage or unleash a hidden atomic bomb (Goudsmit 1996, 133). Perhaps unknown to Goudsmit, the Americans wanted to prevent any hint of the possibility of an atomic bomb to leak out. Another major concern was to keep the scientists out of the hands of the Russians and the French. Similarly, in order to prevent the kidnaping of the captured German rocket scientists, the Americans quickly transported many of them, including Wernher von Braun, to the United States. But with the United States preparing to drop the atomic bombs on Japan, the removal of the German nuclear scientists to the U. S. at such a sensitive time was apparently out of the question.

The guests' treatment derived in part from their legal status. These non-combatant civilian scientists could not be regarded legally as prisoners of war, nor were they charged with any crime. Yet, with the Russians and French eager to snatch them, they could not be left free, even under close supervision. The British had a legal mechanism for such situations, which apparently encouraged the decision for British oversight. According to British law, any person could be "detained at His Majesty's pleasure" without charge for up to six months. On May 26, Max von Laue wrote to his son that he and the others were, quoting the English (with another preposition), "detained *for* his Majesty's pleasure."[30] The expression also appeared in Erich Bagge's diary while at Farm Hall.[31] From then on the prisoners called themselves "the detainees." They remained detainees for exactly six months, from July 3, 1945 to January 3, 1946. The two months of detention on the Continent prior to July 3 did not count in British law.

---

[30]Laue to his son, 26 May 1945, Laue Papers. Italics added.

[31]Bagge's diary, in Bagge et al. (1957, 49).

From *left* to *right*: Heisenberg, Max von Laue, Otto Hahn shortly after their release from Farm Hall (AIP ESVA, Goudsmit Collection)

Several of the detainees were internationally known scientists who had made fundamental discoveries. Two were Nobel Prize winners, Heisenberg and Laue, and one would become a Nobelist at Farm Hall, the chemist Otto Hahn. In addition, Goudsmit was personally acquainted with many of them, and he seemed still to regard them with respect, despite their war-time roles. Probably he also hoped that benevolent treatment would encourage them to reveal any hidden secrets of intelligence interest. Thus, he insisted that they be treated as "guests." Responding in November 1945 to a critic of the scientists' detention at Farm Hall, Goudsmit wrote: "Since the very beginning I have been fighting strenuously to see that these men are treated with special consideration, which is indeed the case... Everything possible is being done to help these guests."[32]

Goudsmit's position contrasted with that at higher levels. As G-2 at SHAEF sought advice from Washington in May 1945 regarding the ultimate disposition of the captive scientists, the Allied Combined Civil Affairs Committee at SHAEF argued for the indefinite detention of all of the scientists and for the complete destruction of German laboratories (Mahoney 1981, 353). It was an approach that many in the U.S. supported. In 1943, U.S. Treasury Secretary Henry Morgenthau, Jr. had presented to President Roosevelt a plan that would impose stringent controls

---

[32]Goudsmit to C.F. Symthe, 26 Nov. 1945, Goudsmit Papers, AIP NBLA, Box 28, Folder 52.

on the entire German population. "Germany's road to peace leads to the farm. The men and women in the German labor force can best serve themselves and the world by cultivating the German soil."[33]

Harvard physicist Edwin C. Kemble (*center*) with two officers in the officers' dining room at the Alsos base in Heidelberg (Michaele Thurgood Haynes and Terry Thurgood, courtesy of NBLA)

In response to the SHAEF recommendation, Goudsmit's second in command, Harvard physicist Edwin C. Kemble, argued that the contribution of German "pure science" (i.e., non-technological) to the war effort was, to the "constant amazement of the scientific consultants of the Alsos mission, actually quite minimal." Moreover, Kemble continued, retaining and fostering these captive established scientists would create "a pool of leadership old enough to recall pre-Hitlerian days and thus [provide] a vital force in the education of postwar German youth."[34] The benevolent treatment of the German scientists won out, dominating American policy as the Cold War set in several years later. The rebuilding of the west German economy, with the support of scientific and technological research, became a priority in order to prevent an economic and cultural vacuum in western Europe. Nevertheless, the revival of German science occurred only under strict Allied

---

[33]H Morgenthau, Jr., "Program to Prevent Germany from Starting World War III," in Morgenthau, *Germany Is Our Problem* (New York: Harper and Brothers, 1945), iii–vi.

[34]Kemble, memo of 9 July 1945, quoted by Mahoney (1981, 353 and 358).

control laws that prevented nuclear and other potentially war-related research (Gimbel 1986).

During their two-month detention before heading to Farm Hall on July 3, the scientific guests moved frequently under Rittner's care through France and Belgium —from Heidelberg to Spa, Reims, Le Vesinet, Huy, then back again. Rittner revealed that one of the reasons for the constant movement was the antagonism of local American officers and the other German prisoners toward the guests for their relative freedom and their higher quality officers' "A rations." These consisted of hot or refrigerated fresh ingredients. Cold, processed "C rations" were the standard fare for prisoners of war. American orderlies at times refused to prepare such fare for the Germans. Nor apparently did local commanders wish to accommodate the German "guests" in comfortable quarters. Rittner reported (as above) that upon their arrival the promised housing was sometimes "no longer available." In addition to preventing any contact between the scientists and their compatriots, avoiding anger and antagonism toward them apparently contributed to the order to Rittner that "no one, repeat no one, was to contact them except on instructions from H. Q." Goudsmit submitted this explanation in response to an inquiry nearly a year later regarding Gerlach: "His detention, and that of a few others, was kept secret for a while because they were treated with special privileges, which would have stirred up a lot of trouble."[35]

Sometime before mid-June 1945 the highest authorities in Washington and London confirmed the decision to remove the detainees from the Continent. They selected the Georgian-style country manor Farm Hall in the village of Godmanchester not far from Cambridge and the Tempsford military air base as their destination. Built in the 1740s as an extension of a much older house, the building and its large grounds had been commandeered during the war for military use. It was returned to private use shortly after the release of the German guests in 1946. It is now occupied by a Cambridge University professor and his wife.

There seemed little other choice but to send the scientists to Britain. The United States and Germany were out of the question, and the security officials would not accept their prolonged internment in France (Mahoney 1981, 373–374). R. V. Jones writes that he was the first to suggest the MI-6 safe house Farm Hall, which was at that time unoccupied and could be used for this purpose. Furthermore, according to Jones, it was he who suggested that microphones should be installed throughout the house and grounds, "so that we could hear their reactions when they realized how far the Americans and ourselves had progressed. If this was an ungentlemanly thing to do, it was a relatively small advantage to be taken of the possible fact that we had saved them from being shot" (Jones 1978, 481–482).

The bugging of German POWs in Britain was, in fact, already standard operating procedure. Helen Fry reports in *The M Room: Secret Listeners Who Bugged the Nazis* (2012) that as early as 1939 the British established an eavesdropping intelligence unit under the command of Lt. Col. Thomas Joseph Kendrick, for the

---

[35]Goudsmit to Noel C. Little, 1 Mar. 1946, Goudsmit Papers, AIP NBLA, Box 28, Folder 51.

purpose of bugging German POWs. The first were held in the Tower of London. Later called MI-19, the unit routinely bugged downed Luftwaffe pilots during the Battle of Britain who were by then held in less threatening and more sumptuous quarters at Trent Park, a country estate north of London. MI-19's biggest operation was the long-term bugging of captured German generals at Trent Park beginning in 1942. They used equipment provided by the American electronics firm RCA and a listening center called the M Room (M presumably for "monitoring"). By the end of the war the listening operation employed 100 listeners and nearly 1000 staff and had expanded to a second facility, Latimer House in Buckinghamshire. Maj. T. H. Rittner was among Kendrick's listeners at Latimer House when he traveled under orders to France in order to take charge of a new group of prisoners, the German scientists. At Rittner's request, Kendrick arranged for technicians and the remainder of the listening team to assemble at Farm Hall. According to Fry, the listening team consisted of Major Rittner, his deputy Capt. P. L. C. Brodie, and six refugee German speakers (Fry 2012, 158 and appendices). Arriving at Farm Hall, the team set up an M Room and installed hidden microphones in an endeavor they code named "Operation 'Epsilon'."

The listeners at Farm Hall. From *left* to *right*: Herbert Lehmann, George Pulay, William Manners, (Leo or Karl Heinz Peter) Kaufmann, Peter Ganz, Maj. T. H. Rittner, (?) Heilbron, Capt. P. L. C. Brodie (Courtesy of Adam Ganz)

Rittner reported in the Preamble to his Farm Hall reports that he paid a visit to Farm Hall shortly before bringing the scientists there.

15th June

Lieut. Cdr. Welsh told me on the telephone that permission had been given for the professors to be brought to England and he asked me to come over as soon as possible to inspect Farm Hall....

17th June

In order to get an air passage to the U.K., I had to get myself temporarily attached to a British unit stationed at BRUSSELS and I accordingly got myself attached to 21 Army Group and got an Authority from them and proceeded to LONDON.

Lieut. Cdr. WELSH and I went to Farm Hall where arrangements had already been made to install microphones. I had asked for such an installation from the day I took charge of the professors. We arranged with Colonel KENDRICK to transfer the necessary staff of technicians from CSDIC (U.K.) [Combined Services Detailed Interrogation Centre] to man the installation. We were fortunate also in obtaining the services of Captain BRODIE from CSDIC (U.K.) to act as Administrative Officer.

26th June

I returned to Belgium leaving Captain BRODIE to complete the arrangements at Farm Hall.

The "professors" (not all held that title) flew to England on July 3 under British guard. They landed at Tempsford airport, and arrived at Farm Hall by car. As the scientists settled into their new quarters, Rittner submitted his first Farm Hall Report, in which he reported the following all the up the chain of command to Gen. Groves:

Microphones have been installed in all the bedrooms and living rooms used by the professors. This installation has proved invaluable as it has enabled us to follow the trend of their thoughts.

In the following conversation, DIEBNER and HEISENBERG discussed the possibility of there being microphones in the house. The conversation took place on 6th July in the presence of a number of their colleagues:

DIEBNER: I wonder whether there are microphones installed here?
HEISENBERG: Microphones installed? (laughing) Oh no, they're not as cute as all that. I don't think they know the real Gestapo methods; they're a bit old fashioned in that respect.

The detainees gave no indication that they suspected otherwise.

The Farm Hall Transcripts

The team of eight agents in the Farm Hall M Room continuously monitored the conversations of their guests and recorded those that appeared to be of intelligence value. They used the latest recording technology provided by RCA: six to eight machines recording onto reusable shellacked metal disks. The selected conversations indicate that the listeners were concerned with such intelligence matters as morale, loyalty to the western allies, political orientation, their mental state, the scientists' reaction to the atomic bombs dropped over Japan, the extent of their knowledge of nuclear fission, the reasons for their lack of progress, and their plans for the future after they returned to Germany.

Rittner and the listeners knew that these were nuclear scientists, but it is not indicated definitively from the transcripts whether or not they knew of the impending atomic bomb. Rittner appeared surprised, but that may have been a ploy to encourage the scientists, especially Hahn, to reveal what he knew. Still, it seems unlikely that he did know about the bomb, since it was a closely held secret. In addition, the Americans and British were eager for the scientists to be surprised, and they probably wanted to prevent Rittner from divulging the information by mistake or the scientists from possibly forcing such information from him. In fact, the scientists did indeed plan one night to get Rittner drunk and then "go for him." But they soon decided against it.

The British agents selected from the recordings only those conversations that would be of special interest to the higher authorities. One listener estimated that the recordings amounted to about 10% of the total conversations (Farm Hall 1993, 12). The agents then translated some or all of the recorded conversations into English, either directly or from written transcriptions. The metal disks were then re-shellacked and reused, thereby destroying the recordings. Maj. Rittner and, after Rittner fell ill, his second-in-command Capt. P. L. C. Brodie, summarized, paraphrased, and excerpted from the transcripts of the conversations, including some important ones in the original German. There were 24 weekly or biweekly reports from Farm Hall during the six-month period of the captivity. Each report was compiled and signed by the officer in charge. For conversations involving technical matters, such as a lecture by Heisenberg on the critical mass of uranium in a bomb, they forwarded the German transcription to London for review and translation. They then included the original and the translation in the final reports. Rittner or Brodie addressed each report to "Mr. M. Perrin and Lt. Cdr. Welsh," but the first copy went to an American Capt. Davis for transmission directly to General Groves; the second went through Perrin and Welsh to British authorities. Each complete set of the reports, numbering 153 pages of standard A4 paper, along with photographs of the detainees and Farm Hall, eventually landed in the archives of their respective nations. Groves's copy is in his papers within the records of the Manhattan Engineer District, National Archives and Records Administration (NARA), College Park, MD, The British copy is at the National Archives, Kew, London and online (see Bibliography).

Several early excerpts from the then-classified Farm Hall Reports appeared in Goudsmit's *Alsos* in 1947, but for security reasons he was not permitted to reveal their source. The existence of these reports remained a secret until 1962, when Groves referred to them in his memoir *Now It Can Be Told* (Groves 1962). He also provided further quotations from which it was evident that the translations were in "workman-like" British English.

Following the publication of Groves's book, many historians, scientists and others urged the American and British governments to declassify and release the full reports to the public. All of these requests met with stubborn refusal, apparently in part because the authorities did not want to reveal the extent of their surveillance, in case it might be needed again. There was also the complicated bureaucratic provenance of the reports, which were classified under the independent auspices of the two governments. In addition, surviving former Farm Hall detainees reportedly objected to their release.

Finally, in 1991, Margaret Gowing, the official historian of the British nuclear program during and after World War II, convinced the presidents of the Royal Society and of the British Academy to submit a letter co-signed by her and 14 other members of these institutions to Lord Mackay of Clashfern, then Lord Chancellor of the government, urging him to approve the release of the Farm Hall Transcripts (Farm Hall 1993, 14–15). The letter assured him that none of the four surviving detainees at the time objected to their release. Lord Mackay responded on February 13, 1992 that the Public Record Office would announce the release of the transcripts the next day (ibid., 16). Ten days later the US National Archives and Records Administration declassified Groves's copies of the transcripts. The American copies were more legible and complete and thus served as the primary basis for the wholly unedited first publication of the reports in Britain in 1993, *Operation Epsilon: The Farm Hall Transcripts* (Farm Hall 1993). In several instances Groves had made markings on the pages, and at least one was made by Goudsmit. In the same year, German physics historian Dieter Hoffmann published a translation of the transcripts back into German (Hoffmann 1993). Three years later American physicist Jeremy Bernstein provided a thoroughly edited and annotated edition, *Hitler's Uranium Club: The Secret Recordings at Farm Hall* (reprint, Bernstein 2001).

Other primary source materials stemming from Farm Hall (and immediately before or after) were already available by the time the transcripts were released, and more appeared after the release. These included Max von Laue's unpublished letters from Farm Hall to his son (see Bibliography), Erich Bagge's diary while at Farm Hall (Bagge et al. 1957), Samuel A. Goudsmit's papers during this entire period (American Institute of Physics, NBLA), and Heisenberg's correspondence with his parents (Heisenberg 2003) and his wife (Heisenberg and Heisenberg 2016, original German 2011). And, for the study of motives and interpretation, we have Niels Bohr's draft letters to Heisenberg as well as Max von Laue's letters to Paul Rosbaud on the scientists' Farm Hall "version [*Lesart*]," quoted earlier. With all of these materials and the complete Farm Hall reports, this highly important, influential and still controversial historical episode, together with its theatrical treatment, is now among the most well documented of that era.

Science, History, Drama

Historical science plays, like historical films and novels, are a form of fiction. Although they are "based on" actual historical events and real science, they deviate of necessity from the actual events (but not from the science, unless it is science fiction) in order to achieve their dramatic impact.

In historical writing an author *tells* his or her readers what happened and how and why it happened. In play writing the author *shows* his or her audience what, how, and why it happened, as she or he chooses to portray it. Most history is written to appeal primarily to the head, to readers' rational faculties; a play is written primarily to appeal to the heart, to viewers' feelings and emotions, the foundations of drama. But neither is exclusive. Because of their differences, history incorporated into drama must become, to some extent, fictionalized in order to achieve the dramatic aims. At the same time, the overall drama must remain true to the general features of the history if it is not to devolve entirely into fiction.

I encountered these distinctions while working on the early drafts of *Farm Hall*, my first play, and I had much to learn. As a science historian, I had spent years trying to get the facts straight: to get the exact quotations, the exact sequence of events, the exact documentary basis underlying the events. Now I had to abandon much of that goal for the sake of another goal: encompassing the human, emotional content of the events in order to make those events come alive on a stage before a live audience. As one actor informed me, "It's not history; it's theatre!" It was a very rewarding experience, for it enabled me, and audiences, to experience for the first time the historical events and characters as actual, real events and people.

Real life is normally not very dramatic, nor are most events of great significance. The early drafts of *Farm Hall* displayed this well. When I showed them to my most insightful critic, my spouse, she responded: "But it's just a bunch of men sitting

© Springer International Publishing AG 2017   89
D.C. Cassidy, *Farm Hall and the German Atomic Project of World War II*,
DOI 10.1007/978-3-319-59578-8_3

around talking!" She was right. The historical reality had not yet come alive as human drama. Through the advice of patient dramatists and the reading and viewing of great similar dramas, such as *Twelve Angry Men*, *Conspiracy*, and *Stalag 17*, I gradually realized that, even if it is a bunch of men sitting around talking, the dramatic action of a play exists instead within the heightened tensions and emotions flying about the room—in the fears, anxieties, conflicts, failures, and difficult reactions to each other and to the momentous events occurring outside their little world.

University of Oxford literature professor Kirsten Shepherd-Barr, who has written probably the only book-length analysis so far of science plays (Shepherd-Barr 2006), discusses the dramatic tension the historical playwright faces between fiction and non-fiction as follows: "The playwright wants to have it both ways: to preserve absolute authenticity and fidelity to the 'data' and 'facts,' yet still manipulate them in the name of theatricality and drama." In the end, she argues, drama will always prevail over history because of the very nature of a play. She uses as an example Heinar Kipphardt's famous play *In the Matter of J. Robert Oppenheimer* which, much like *Farm Hall*, was based on a trove of recorded conversations. In his case it was the verbatim testimonies given during Oppenheimer's security hearings.

> One could argue that even if he did not invent any dialogue, even if the lines were verbatim as stated in the proceedings, the mere act of placing these factual texts in the mouths of actors and putting them on a stage alters the historical authenticity of the experience, because it introduces the element of mediation that performance involves. The audience knows it is watching a play, and that no matter how authentic that play claims to be, it is an interpretation, not a re-creation, of real people and events. (Shepherd-Barr 2006, 186–187)

But how far is a playwright permitted to bend the truth, the actual facts of an event, in the interest of dramatic success? As with much else in Western drama, I found the best answer to that question, at least for me, in the works of those who invented the genre, the authors of classic Greek tragedy and epic poetry, and in particular in Aristotle's *Poetics*. His analysis of this art form has served as a foundation of Western drama ever since, although, of course, the dramatic form has undergone much further development through the works of Shakespeare, Goethe, Chekhov, O'Neill and many other influential dramatists over the centuries. Nearly two-and-a-half millennia ago Aristotle wrote regarding history and epic poetry, both forms of story-telling:

> From what we have said it will be seen that the poet's function is to describe, not the thing that has happened, but a kind of thing that might happen, i.e. what is possible as being probable or necessary. The distinction between historian and poet is not in the one writing prose and the other verse…it consists really in this, that the one describes the thing that has been, and the other a kind of thing that might be. Hence poetry is something more philosophic and of graver import than history, since its statements are of the nature rather of universals, whereas those of history are singular. (Aristotle 1984, 2322–2323)

As literature critic Eric Bentley put it, following Georg Lukács, "while playwrights and novelists depart from the *facts* of history, they still present the larger *forces* of history" (Bentley 1952, 13).

Closely associated with the issue of historical accuracy is the issue of justice. Audiences generally want justice to prevail and are uneasy when it does not. This issue has arisen in the case of *Farm Hall*. Some audience members expressed disappointment and even anger that the German scientists who had worked on the potentially devastating field of nuclear fission under the Hitler regime were treated so elegantly at Farm Hall and then went on to such highly successful post-war careers. Whether or not one sees this as unjust, the play does highlight the scientists' privileged situation compared with that of their compatriots back home, and, in the last scene, their highly successful post-war careers. In this way the play also serves as a demonstration that, in some frustrating instances, real life is a "thing" that cannot be bent beyond breaking. That some audience members expressed an emotional response to this was one indication, my drama colleagues assured me, that the play was a success.

One of the unique features of a science- history play is the need to deal with the science, at least to some extent. But science is a subject that most general audiences prefer to avoid if at all possible. The great popularity of Michael Frayn's play *Copenhagen* was for me an inspiration and a lesson in how to incorporate science successfully without losing the audience. One of the surprising aspects of that remarkable play is the large amount of science that it does contain, yet audiences eagerly accepted it and emerged from the play pleased that they had done so. One way to do this, the play shows, is to present the science in such a way that the audience does not realize that it is actually learning science; or if it does, it *wants* to learn it. And one way to accomplish that is to render the science essential to the dramatic flow of the play. In order to engage with that flow, the audience wants to be attentive when, for instance, one character explains the basic ideas to another character. Of course, the explanation must be brief and easily accessible to the intended audience. In addition, Shepherd-Barr finds that the best science plays go beyond explanation by building a scientific principle from the play–such as uncertainty, relativity, the attraction or repulsion of likes or opposites—into the structure of the play itself.

Shepherd-Barr refers to historical plays that adhere as closely as possible to the details of the actual events "docudramas," as opposed to documentaries, a designation that fits *Farm Hall* well (Shepherd-Barr 2006, 63). Because of the close association of the drama with the history in this play, many people have asked how closely the characters, dialogue, quoted letters, and events in the play reflect their actual counterparts. In order to answer this question, and to provide material for further study of both the events and the play-writing process, I discuss below some of the various materials and how they were used in creating the play.

The Farm Hall reports, with their extensive transcriptions of the scientists' conversations at Farm Hall in English translation, naturally comprised the principal source for the play and are discussed in a separate section. Likewise, Rittner's commentary in his reports rendered him the natural character to serve as narrator. The reports were especially useful for the second half of the play, after the dropping of the first atomic bomb on Japan. But one major deviation from the facts is immediately obvious: While there were ten German captives at Farm Hall, there are only five captives in *Farm Hall*. The reason for this is the practical difficulty for the playwright, the audience, and the director of managing all ten personalities on stage, and for the theater company of engaging ten actors plus the other characters in a full production. The differences among the ten are not outstanding, while five characters enable a much clearer depiction of the differing personalities, positions, and conflicts. I also discovered that five is an optimal number for building dramatic conflict, because it enables the formation of two opposing alliances of two characters each with one "floating" character. In this case, Heisenberg and Weizsäcker are roughly allied against Gerlach and Diebner, with Hahn as the "floater." But it is Diebner and Heisenberg who exhibit the sharpest conflicts among them.

Compared with the second half of the play, the pre-bomb scenes deviated more from the actual transcripts. This occurred because of the need to invent more of the "possible" or "probable" in order to enhance the sudden dramatic contradictions that will occur with the dropping of the bomb. Those scenes also had to convey a great deal of needed background information about who these scientists are; their current location and situation; how they rationalized their work and allegiances during the war; the emphasis on their mistaken belief in the continued superiority and progress of the German effort in comparison to that of the Allies; the establishment of tensions and alliances among these characters; and their very human fears, concerns and worries in their situation. Drawing upon my prior work and utilizing other historical sources, those scenes were also designed to reveal more of Heisenberg's human side in order to render him vulnerable and sympathetic, along with his obvious flaws, and thus to encourage the audience to engage more with him and with the play itself. This was done partly though the letters to and from his wife (who also adds a female presence to the play) and partly through the two revelations of his failings during the first half of the play: one at the very outset regarding Goudsmit's parents; the other in his admission of failure in effectively counteracting the Nazi policies. He will have to admit further failures regarding the critical mass and the meager results of his work in comparison with the Allied success. All of this was intended to culminate in the parallels with Goudsmit and the "moral" of the story expressed by Rittner and Goudsmit at the end of Scene 15.

Samuel A. Goudsmit (AIP ESVA)

The play also draws upon Samuel A. Goudsmit's book *Alsos*, which contains an account of his initial interrogation of Heisenberg (pp. 112–113) as inspiration for Scene 1; as well as Goudsmit's return (actually alone) to his devastated home in The Hague (pp. 46–49) and the photograph incident (pp. 98–99), both of which were inspirations for Scene 15. The photograph incident also appeared in the archival records of the Alsos Mission (see previous chapter). Heisenberg's correspondence with his wife in Scenes 5, 7, and 14 are somewhat embellished composites of their letters published in Heisenberg and Heisenberg (2016) (original German 2011). Elisabeth's letter to Heisenberg on August 20, 1945, used in Scene 7, was also based on Rittner's report of it in Farm Hall Report 6 (Farm Hall 1993, pp. 170–171).

The statement by the Voice of Doom in Scene 10 has been attributed to J. Robert Oppenheimer. He was reportedly quoting from the *Bhagavad Gita* upon seeing the first nuclear detonation, on July 16, 1945 at the Trinity site in New Mexico.

Rittner's charge to Goudsmit in Scene 2 is derived from Rittner's report of his activities on May 1, 1945 in the preamble to his reports. Since very little is known about Rittner, I had to invent most of his "possible" life story and background. The complete BBC report excerpted in Scene 11 may be found in Bernstein (2001, 357–358). The newspaper report in Scene 12 is from the *London Times*, August 7, 1945, which contained an excerpt from former Prime Minister Winston Churchill's statement about the bomb that he gave during a BBC broadcast the day before.

Historical Sources: The Farm Hall Reports

Presented below are extended excerpts from the Farm Hall reports that inspired many of the scenes in the play, as well as an excerpt from Erich Bagge's diary written while at Farm Hall. The Farm Hall texts are derived from *Operation Epsilon: The Farm Hall Transcripts* (Farm Hall 1993), which closely followed the original, unedited format of the reports. The unedited format is retained here. A partial line of ellipses indicates excluded material. The reports are written in British English and in standard military style, with numbered sections and paragraphs and capitalized names of people and places for easy reference. Farm Hall Report 4, on the scientists' reactions to the atom bomb dropped on Hiroshima, is especially significant. The excerpt from Report 18 at the end, regarding the award of the Nobel Prize to Otto Hahn, is followed by Bagge's fuller account of the impact of the news of the award in his diary. This became the main inspiration for Scene 13.

Additional excerpts from the Farm Hall reports, including the Preamble, may be found earlier in this book. A fully edited and annotated edition of the reports is available in Bernstein 2001 and 1996. The original reports are on deposit in the American and British national archives (see Bibliography).

© Springer International Publishing AG 2017   95
D.C. Cassidy, *Farm Hall and the German Atomic Project of World War II*,
DOI 10.1007/978-3-319-59578-8_4

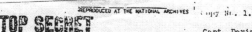

TOP SECRET

Copy No. 1.

Capt. Davis
For General Groves.

Ref: F.H./5.

To: Mr. M. PERRIN and Lt. Cdr. WELSH.
From: Major T.H. RITTNER.

OPERATION EPSILON.

(8th-22nd August, 1945).

I.   GENERAL.

1.   The guests have recovered from the initial shock
they received when the news of the atomic bomb was
announced. They are still speculating on the method
used to make the bomb and their conversations on this
subject, including a lecture by HEISENBERG, appear
later in this report. The translation of the technical
matter has been very kindly undertaken by a member of
the Staff of D.S.I.R. The original German text of
HEISENBERG's lecture has been reproduced as an appendix
to this report.

2.   There is a general air of expectancy as the guests
now feel there is no further need for their detention
and they assume that they will shortly be told what plans
have been made for their future and that they will soon
be reunited with their families. They are eagerly awaiting replies to their letters which have now been despatched.

3.   The declaration of the surrender of JAPAN was
greeted with relief rather than enthusiasm. The guests listened with great interest to the King's broadcast on "VJ"
Day and all stood rigidly to attention during the playing
of the National Anthem.

4.   Sir CHARLES DARWIN paid a visit to FARM HALL
on 18th August. This was the first time the guests had
had contact with a scientist since their detention and
they were delighted to have the opportunity of meeting
him. Conversations during the visit and subsequent
reactions are dealt with elsewhere in this report.

II.   The Future.

1.   A number of the guests have discussed their
attitude towards co-operation with the Allies. The
following conversation took place between HEISENBERG,
VON WEISZACKER and GERLACH on 10th August.

GERLACH:
        If you were faced with the opportunity of co-
operation in order to make the bomb useful for mankind,
would you do it?

First page of Farm Hall Report 5 (NARA, Farm Hall Reports)

REPRODUCED AT THE NATIONAL ARCHIVES   Copy No. 1.

TOP SECRET

- 4 -

HAHN:
DOEPEL was the first to discover the increase in neutrons.

HARTECK:
Who is to blame.

(?) VOICE:
HAHN is to blame.

WEIZSACKER:
I think it's dreadful of the Americans to have done it. I think it is madness on their part.

HEISENBERG:
One can't say that. One could equally well say "That's the quickest way of ending the war.

HAHN:
That's what consoles me.

HEISENBERG:
I still don't believe a word about the bomb but I may be wrong. I consider it perfectly possible that they have about ten tons of enriched uranium, but not that they can have ten tons of pure U. 235.

HAHN:
I thought that one needed only very little 235.

HEISENBERG:
If they only enrich it slightly, they can build an engine which will go but with that they can't make an explosive which will.-

HAHN:
But if they have, let us say, 30 kilogrammes of pure 235, couldn't they make a bomb with it?

HEISENBERG:
But it still wouldn't go off, as the mean free path is still too big.

HAHN:
But tell me why you used to tell me that one needed 50 kilogrammes of 235 in order to do anything. Now you say one needs two tons.

HEISENBERG:
I wouldn't like to commit myself for the moment, but it is certainly a fact that the mean free paths are pretty big.

HARTECK:
Do you want 4 or 5 centimetres, - then it would break up on the first or second collision.

HEISENBERG:
But it needn't have the diameter of only 4 or 5 centimetres.

DECLASSIFIED
UK Note 24/28 92
By JS/mcA, NARS, Date 2/25/92

Page 4 from Farm Hall Report 5 (NARA, Farm Hall Reports)

Excerpts from the Farm Hall Reports

From FARM HALL REPORT 1

COPY NO.1
Mr. M. Perrin for General Groves,
through Lt. Cdr. E. Welsh
Ref. FH1 [Farm Hall 1]

To: Mr. M. Perrin and Lt. Comdr. Welsh.
From: Major T.H. Rittner.

OPERATION "EPSILON"
(3rd–18th July 45)

I. General

1. A report covering the operation on the continent from May 2nd until 3rd July 1945 has already been submitted.
2. The arrangements for bringing the party to England\vent according to plan and the following landed at TEMPSFORD on the afternoon of 3rd July and were taken to FARM HALL by car.

Professor VON LAUE.
Professor HAHN.
Professor HEISENBERG.
Professor GERLACH.
Doctor HARTECK.
Doctor VON WEIZSACKER.
Doctor WIRTZ.
Doctor DIEBNER.
Doctor BAGGE.
Doctor KORSCHING

together with four PW orderlies. A further PW orderly has since been added to the party.

3. All the Professors have renewed their parole to me in writing in respect of FARM HALL and grounds and I have warned them that any attempt by any one of them or by the orderlies to escape or to communicate with anyone will result in them all having their liberty considerably restricted.
4. Ordinary army rations are drawn for the professors and the officers and troops, and these are prepared for all by the PW cooks.
5. Microphones have been installed in all the bedrooms and living rooms used by the professors. This installation has proved invaluable as it has enabled us to follow the trend of their thoughts

In the following conversation, DIEBNER and HEISENBERG discussed the possibility of there being microphones in the house. The conversation took place on 6th July in the presence of a number of their colleagues:

DIEBNER: I wonder whether there are microphones installed here?

HEISENBERG: Microphones installed? (*laughing*) Oh no, they're not as cute as all that. I don't think they know the real Gestapo methods; they're a bit old fashioned in that respect.

II. MORALE

1. The party has settled down well at FARM HALL but they are becoming more and more restive. The question of their families is causing them the greatest anxiety and I believe that if it were possible to make arrangements for an exchange of messages with their families, the effect on general morale would be immediate.

2. Most of the recorded conversations are of a general nature and show that they are pleased with the treatment they are receiving but completely mystified about their future.

…

(a) Conversation between HEISENBERG. HARTECK, WIRTZ, DIEBNER and KORSCHING after the announcement of Lt. Comdr. WELSH's visit:

HEISENBERG: I can see the time is coming when we must have a very serious talk with the Commander. Things can't go on like this.

HARTECK: It won't do. We have no legal position since they have to keep us hidden.

…

HEISENBERG: I should say that the point is that they don't yet know what they want. That's the whole trouble. They don't want us to take part in any discussion regarding our future as they don't want us to have any say in the matter. They want to consider what to do and they have not yet agreed among themselves.

…

From FARM HALL REPORT 2

OPERATION "EPSILON"

(18–31 July 45)

I. General

There has been very little change in the position at FARM HAL since the last report. Outwardly the guests are serene and calm, but it is clear that their restiveness is increasing. Suggestions have been made that one of the guests should attempt to get a letter to CAMBRIDGE. Steps have been taken to prevent this.

…

HEISENBERG: That's all much too pessimistic. I think there is a 35% chance that
               we will get back before 1 December. The chance of our getting
               back within a reasonable time after that date I would put at 50%.
               The chance of our never getting back except perhaps in totally
               different circumstances after many, many years, I would put at
               14%. There is a 1% chance that we will never see our wives
               again. I can see no reason to assume that they want to treat us
               badly, but I can see a reason to assume that they don't want to
               have us in Germany as they don t want us to pass on our
               knowledge to other people.

HARTECK: That is one point but on the other hand we may be shot; not by
               the English but by the people there. If one of us went to Hamburg
               University some mad student might come and shoot one.

...

5. HAHN and DIEBNER had a long talk on 30 July part of which is reproduced
   elsewhere in this report. The following extract shows their attitude to the letter
   question and HAHN's philosophical acceptance of the situation:

HAHN: I read an article in the Picture Post about the Uranium bomb: it said
               that the newspapers had mentioned that such a bomb was being made
               in Germany. Now you can understand that we are being "detained"
               because we are such men. They will not let us go until they are
               absolutely certain that no harm can be done or that we will not fall
               into Russian hands or anything like that. To my mind it is a mistake to
               do anything. All my hopes and efforts are now directed towards
               getting into touch with my family... The longer one is "detained" here
               and knows nothing, the more one gets into a state where one racks
               one's brain to discover what is going to happen. I fight against it and
               make jokes. Also I don't take life too seriously in that I always look
               on the bright side of things.

DIEBNER: I would have been just the same in Germany. The day before I went
               away I said to my wife: "I suggest we commit suicide". I had reached
               that stage then.

HAHN: My wife was like that sometimes and that is why I am worried
               whether she will hold out without news. See what LAUE did against
               National Socialism and I think I worked against it too. We are both
               innocent but I am not allowed to write to my wife. I have told the
               Major: "If my American and English friends knew how I am being
               repaid for all my work since 1933, that I am not even allowed to write
               my wife, they would be very surprised". We are being well treated
               here, our slightest wish is granted if it is possible, everything except
               writing letters.

...

2. Part of a conversation between DIEBNER and HAHN on 30 July:

DIEBNER: I wanted to tell you how I came to join the Party and how I have suffered under the Nazis. In 1933 I became a Freemason in opposition to National Socialism. I <u>never</u> voted for Hitler. That became known in HALLE and the result was that I got into difficulties at the Institute. Then I went to the "Waffenamt" and was to have become a civil servant, but I did not. SCHUMANN didn't forward my application. He said he couldn't do it because I was a Freemason. SCHUMANN did his best for me and sent me to a man in Munich and after a year the thing went through and I became a civil servant, a "Regierungsrat".

HAHN: The fact of being a Party member does not necessarily tell against a man. The newspapers say that.... The fact that you were in the Party hasn't really done you any harm.

...

IV. The Future

Speculation by the guests as to the future in general has been dealt with under the heading "Morale", but the following conversation between DIEBNER. KORSCHING and BAGGE on 21 Jul. goes rather further:

BAGGE: For the sake of the money. I should like to work on the Uranium engine; on the other hand, I should like to work on cosmic rays. I feel like DIEBNER about that.

KORSCHING: Would you both like to construct a Uranium-engine.?

DIEBNER: That is <u>the</u> chance to earn a living.

...

BAGGE: I am convinced they (Anglo-Americans) have used these last 3 months mainly to imitate our experiments.

KORSCHING: Not even that. They used them to discuss with their experts their possibilities and to study the secret documents. They probably examined a few specimens of our Uranium-blocks. From these specimens they can see for instance whether the engine has been running already. It could have been run; the blocks must have undergone some internal chemical change.

BAGGE: But they know already, that it did not run; that they were told.

KORSCHING: That is just it. They were told practically everything up to approximately the last series of measurements. It is the same to them whether it ever came to an increase in Neutrons of 5 or 50. The issue must be quite clear to them.

BAGGE: But they'll certainly have the ambition to imitate our experiments as soon as possible and for that purpose they need the D20. Once they have worked with that—(int.)

...

From FARM HALL REPORT 3

OPERATION "EPSILON"

(1–6 August, 1945)

...

II. Morale

In conversation with a British Officer regarding the position of communication with the families, HAHN completely broke down. BAGGE also came very near to tears when he described the fate worse than death which he pictured was that of his wife and children at the hands of the Moroccan troops.

General morale has however improved since I was able to tell the guests that permission had been granted for them to write letters to their families and that it was hoped to obtain answers. This permission was contained in a cable from Lt. Cdr. WELSH to Mr. PERRIN dated 1 August.

...

From FARM HALL REPORT 4

OPERATION "EPSILON"

(6–7th August, 1945)

I. Preamble

1. This report covers the first reactions of the guests to the news that an atomic bomb had been perfected and used by the Allies.
2. The guests were completely staggered by the news. At first they refused to believe it and felt that it was bluff on our part to induce the Japanese to surrender. After hearing the official announcement, they realised that it was a fact. Their first reaction, which I believe was genuine, was an expression of horror that we should have used this invention for destruction.
3. The appendices to this report are:

    1. Declaration signed by all the guests setting out details of the work in which they were engaged in Germany.
    2. Photographs of the guests with brief character sketches of each.

II. 6th August, 1945

1. Shortly before dinner on the 6th August I informed Professor HAHN that an announcement had been made by the B.B.C. that an atomic bomb had been dropped. HAHN was completely shattered by the news and said that he felt personally responsible for the deaths of hundreds of thousands of people, as it was his original discovery which had made the bomb possible. He told me that he had originally contemplated suicide when he realised the terrible potentialities of his discovery and he felt that now these had been realised and he was to blame. With the help of considerable alcoholic stimulant, he was calmed down and we went down to dinner where he announced the news to the assembled guests.

2. As was to be expected, the announcement was greeted with incredulity. The following is a transcription of the conversation during dinner.

HAHN:	They can only have done that if they have uranium isotope separation.
WIRTZ:	They have it too.
HAHN:	I remember SEGRE's, DUNNING's and my assistant GROSSE's work: they had separated a fraction of a milligramme before the war, in 1939.
LAUE:	235?
HAHN:	Yes. 235.
HARTECK:	That's not absolutely necessary. If they let a uranium engine run, they separate "93" [neptunium].
HAHN:	For that they must have an engine which can make sufficient quantities of "93" to be weighed.
GERLACH:	If they want to get that, they must use a whole ton.
HAHN:	An extremely complicated business, for "93" they must have an engine which will run for a long time. If the Americans have a uranium bomb then you're all second-raters. Poor old HEISENBERG.
LAUE:	The innocent!
HEISENBERG:	Did they use the word uranium in connection with this atomic bomb?
ALL:	No.
HEISENBERG:	Then it's got nothing to do with atoms. but the equivalent of 20,000 tons of high explosive is terrific.
WEIZSACKER:	It corresponds exactly to the factor 10^4.
GERLACH:	Would it be possible that they have got an engine running fairly well, that they have had it long enough to separate 93?
HAHN:	I don't believe it.
HEISENBERG:	All I can suggest is that some dilettante in America who knows very little about it has bluffed them in saying "If you drop this it has the equivalent of 20,000 tons of high explosive" and in reality doesn't work at all.
HAHN:	At any rate HEISENBERG you're just second-raters and you may as well pack up.
HEISENBERG:	I quite agree.
HAHN:	They are fifty years further advanced than we.
HEISENBERG:	I don't believe a word of the whole thing. They must have spent the whole of their £500,000,000 in separating isotopes: and then it's possible.
WEIZSACKER:	If it's easy and the Allies know it's easy, then they know that we will soon find out how to do it if we go on working.
HAHN:	I didn't think it would be possible for another twenty years.
WEIZSACKER:	I don't think it has anything to do with uranium.

HAHN: It must have been a comparatively small atomic bomb—a hand one.

HEISENBERG: I am willing to believe that it is a high pressure bomb and I don't believe that it has anything to do with uranium but that it is a chemical thing where they have enormously increased the speed of the reaction and enormously increased the whole explosion.

GERLACH: They have got "93" and have been separating it for two years. somehow stabilised it at low temperature and separated "93" continuously.

HAHN: But you need the engine for that.

DIEBNER: We always thought we would need two years for one bomb.

HAHN: If they have really got it, they have been very clever in keeping it secret.

WIRTZ: I'm glad we didn't have it.

WEIZSACKER: That's another matter. How surprised BENZER (?) (MENZEL? SAG [Goudsmit]) would have been. They always looked upon it as a conjuring trick.

WIRTZ: DOEPEL, BENZER (?) and Company.

HAHN: DOEPEL was the first to discover the increase in neutrons.

HARTECK: Who is to blame?

(?) VOICE: HAHN is to blame.

WEIZSACKER: I think it's dreadful of the Americans to have done it. I think it is madness on their part.

HEISENBERG: One can't say that. One could equally well say "That's the quickest way of ending the war".

HAHN: That's what consoles me.

HEISENBERG: I still don't believe a word about the bomb but I may be wrong. I consider it perfectly possible that they have about ten tons of enriched uranium, but not that they can have ten tons of pure U.235.

HAHN: I thought that one needed only very little 235.

HEISENBERG: If they only enrich it slightly, they can build an engine which will go but with that they can't make an explosive which will—

HAHN: But if they have, let us say, 30 kg of pure 235, couldn't they make a bomb with it?

HEISENBERG: But it still wouldn't go off, as the mean free path is still too big.

HAHN: But tell me why you used to tell me that one needed 50 kg of 235 in order to do anything. Now you say one needs two tons.

HEISENBERG: I wouldn't like to commit myself for the moment, but it is certainly a fact that the mean free paths are pretty big.

HARTECK: Do you want 4 or 5 cm.—then it would break up on the first or second collision.

HEISENBERG:	But it needn't have the diameter of only 4 or 5 cm.
HAHN:	I think it's absolutely impossible to produce one ton of uranium 235 by separating isotopes.
WEIZSACKER:	What do you do with these centrifuges?
HARTECK:	You can never get pure 235 with the centrifuge. But I don't believe that it can be done with the … centrifuge.
WIRTZ:	No, certainly not.
HAHN:	Yes. but they could do it too with the mass-spectrographs. EWALD has some patent.
DIEBNER:	There is also a photo-chemical process.
HEISENBERG:	There are so many possibilities, but there are none that we know. that's certain.
WIRTZ:	None which we tried out.
HAHN:	I was consoled when, I believe it was WEIZSACKER, said that there was now this uranium—23 min—I found that in my institute too, this absorbing body which made the thing impossible consoled me because when they said at one time one could make bombs, I was shattered.
WEIZSACKER:	I would say that at the rate we were going, we would not have succeeded during this war.
HAHN:	Yes.
WEIZSACKER:	It is very cold comfort to think that one is personally in a position to do what other people would be able to do one day.
HAHN:	Once I wanted to suggest that all uranium should be sunk to the bottom of the ocean. I always thought that one could only make a bomb of such a size that a whole province would be blown up.
HEISENBERG:	If it has been done with uranium 235 then we should be able to work it out properly. It just depends upon whether it is done with 50, 500 or 5000 kg and we don't know the order of magnitude. We can assume that they have some method of separating isotopes of which we have no idea.
WIRTZ:	I would bet that it is a separation by diffusion with recycling.
HEISENBERG:	Yes, but it is certain that no apparatus of that sort has ever separated isotopes before. KORSCHING might have been able to separate a few more isotopes with his apparatus.
WIRTZ:	We only had one man working on it and they may have had ten thousand.
WEIZSACKER:	Do you think it is impossible that they were able to get element "93" or "94" out of one or more running engines?
WIRTZ:	I don't think that is very likely.

WEIZSACKER: I think the separation of isotopes is more likely because of the interest which they showed in it to us and the little interest they showed for the other things.

HAHN: Well, I think we'll bet on HEISENBERG's suggestion that it is a bluff.

...

3. All the guests assembled to hear the official announcement at 9 o'clock. They were completely stunned when they realised that the news was genuine. They were left alone on the assumption that they would discuss the position and the following remarks were made:

HARTECK: They have managed it either with mass-spectrographs on a large scale or else they have been successful with a photo-chemical process.

WIRTZ: Well I would say photo-chemistry or diffusion. Ordinary diffusion. They irradiate it with a particular wave-length—(all talking together).

HARTECK: Or using mass spectrographs in enormous quantities. It is perhaps possible for a mass-spectrograph to make one milligramme in one day—say of "235". They could make quite a cheap mass-spectrograph which in very large quantities might cost a hundred dollars. You could do it with a hundred thousand mass-spectrographs.

HEISENBERG: Yes. of course. if you do it like that: and they seem to have worked on that scale. 180,000 people were working on it.

HARTECK: Which is a hundred times more than we had.

BAGGE: GOUDSMiT led us up the garden path.

HEISENBERG: Yes. he did that very cleverly.

HAHN: CHADWICK and COCKROFT.

HARTECK: And SIMON too. He is the low temperature man.

KORSCHING: That shows at any rate that the Americans are capable of real cooperation on a tremendous scale. That would have been impossible in Germany. Each one said that the other was unimportant.

GERLACH: You really can't say that as far as the uranium group is concerned. You can't imagine any greater cooperation and trust than there was in that group. You can't say that any one of them said that the other was unimportant.

KORSCHING: Not officially of course.

GERLACH: (Shouting). Not unofficially either. Don't contradict me. There are far too many other people here who know.

HAHN: Of course we were unable to work on that scale.

HEISENBERG:	One can say that the first time large funds were made available in Germany was in the spring of 1942 after that meeting with RUST when we convinced him that we had absolutely definite proof that it could be done.
BAGGE:	It wasn't much earlier here either.
HARTECK:	We really knew earlier that it could be done if we could get enough material. Take the heavy water. There were three methods, the most expensive of which cost 2 marks per gramme and the cheapest perhaps 50 pfennigs. And then they kept on arguing as to what to do because no one was prepared to spend 10 millions if it could be done for three millions.
HEISENBERG:	On the other hand, the whole heavy water business which I did everything I could to further cannot produce an explosive.
HARTECK:	Not until the engine is running.
HAHN:	They seem to have made an explosive before making the engine and now they say: "in future we will build engines".
HARTECK:	If it is a fact that an explosive can be produced either by means of the mass spectrograph—we would never have done it as we could never have employed 56,000 workmen. For instance, when we considered the CLUSIUS-LINDE business combined with our exchange cycle we would have needed to employ 50 workmen continuously in order to produce two tons a year. If we wanted to make ten tons we would have had to employ 250 men. We couldn't do that.
WEIZSACKER:	How many people were working on V 1 and V 2?
DIEBNER:	Thousands worked on that.
HEISENBERG:	We wouldn't have had the moral courage to recommend to the Government in the spring of 1942 that they should employ 120,000 men just for building the thing up.
WEIZSACKER:	I believe the reason we didn't do it was because all the physicists didn't want to do it, on principle. If we had all wanted Germany to win the war, we would have succeeded.
HAHN:	I don't believe that but I am thankful we didn't succeed.
HARTECK:	Considering the figures involved I think it must have been mass spectrographs. If they had had some other good method, they wouldn't have needed to spend so much. One wouldn't have needed so many men.
WIRTZ:	Assuming it was the CLUSIUS method they would never have been able to do anything with gas at high temperatures.
HARTECK:	When one thinks how long it took for us to get the nickel separating tube I believe it took nine months.
KORSCHING:	It was never done with spectrographs.
HEISENBERG:	I must say I think your theory is right and that it is spectrographs.
WIRTZ:	I am prepared to bet that it isn't.

HEISENBERG: What would one want 60,000 men for?

KORSCHING: You try and vaporise one ton of uranium.

HARTECK: You only need ten men for that. I was amazed at what I saw at I.G. [Farben].

HEISENBERG: It is possible that the war will be over tomorrow.

HARTECK: The following day we will go home.

KORSCHING: We will never go home again.

HARTECK: If we had worked on an even larger scale we would have been killed by the "Secret Service". Let's be glad that we are still alive. Let us celebrate this evening in that spirit.

DIEBNER: Professor GERLACH would be an Obergruppenfuhrer and would be sitting in LUXEMBOURG as a war criminal.

KORSCHING: If one hasn't got the courage. it is better to give up straightaway.

GERLACH: Don't always make such aggressive remarks.

KORSCHING: The Americans could do it better than we could, that's clear.

                (GERLACH leaves the room.)

HEISENBERG: The point is that the whole structure of the relationship between the scientist and the state in Germany was such that although we were not 100% anxious to do it, on the other hand we were so little trusted by the state that even if we had wanted to do it it would not have been easy to get it through.

DIEBNER: Because the official people were only interested in immediate results. They didn't want to work on a long-term policy as America did.

WEIZSACKER: Even if we had got everything that we wanted, it is by no means certain whether we would have got as far as the Americans and the English have now. It is not a question that we were very nearly as far as they were but it is a fact that we were all convinced that the thing could not be completed during this war.

HEISENBERG: Well that's not quite right. I would say that I was absolutely convinced of the possibility of our making an uranium engine but I never thought that we would make a bomb and at the bottom of my heart I was really glad that it was to be an engine and not a bomb. I must admit that.

WEIZSACKER: If you had wanted to make a bomb we would probably have concentrated more on the separation of isotopes and less on heavy water.

                (HAHN leaves the room)

WEIZSACKER: If we had started this business soon enough we could have got somewhere. If they were able to complete it in the summer of 1945, we might have had the luck to complete it in the winter 1944/45.

WIRTZ:	The result would have been that we would have obliterated LONDON but would still not have conquered the world, and then they would have dropped them on us.
WEIZSACKER:	I don't think we ought to make excuses now because we did not succeed, but we must admit that we didn't want to succeed. If we had put the same energy into it as the Americans and had wanted it as they did, it is quite certain that we would not have succeeded as they would have smashed up the factories.
DIEBNER:	Of course they were watching us all the time.
WEIZSACKER:	One can say it might have been a much greater tragedy for the world if Germany had had the uranium bomb. Just imagine. if we had destroyed LONDON with uranium bombs it would not have ended the war, and when the war did end, it is still doubtful whether it would have been a good thing.
WIRTZ:	We hadn't got enough uranium.
WEIZSACKER:	We would have had to equip long distance aircraft with uranium engines to carry out airborne landings in the CONGO or NORTH WEST CANADA. We would have had to have held these areas by military force and produce the stuff from mines. That would have been impossible.
HARTECK:	The uranium content in the stone in the radium mines near GASTEIN was said to be so great that the question of price does not come into it.
BAGGE:	There must be enormous quantities of uranium in UPPER SILESIA. Mining experts have told me that.
DIEBNER:	Those are quite small quantities.
HARTECK:	If they have done it with mass-spectrographs, we cannot be blamed. We couldn't do that. But if they have done it through a trick, that would annoy me.
HEISENBERG:	I think we ought to avoid squabbling amongst ourselves concerning a lost cause. In addition, we must not make things too difficult for HAHN.
HARTECK:	We have probably considered a lot of things which the others cannot do and could use.
WEIZSACKER:	It is a frightful position for HAHN. He really did do it.
HEISENBERG:	Yes. (Pause) About a year ago, I heard from SEGNER(?) from the Foreign Office that the Americans had threatened to drop a uranium bomb on Dresden if we didn't surrender soon. At that time I was asked whether I thought it possible, and, with complete conviction, I replied: "No".
WIRTZ:	I think it characteristic that the Germans made the discovery and didn't use it, whereas the Americans have used it. I must say I didn't think the Americans would dare to use it.

4. HAHN and LAUE discussed the situation together. HAHN described the news as a tremendous achievement without parallel in history and LAUE expressed the hope of speedy release from detention in the light of these new events.

5. When GERLACH left the room he went straight to his bedroom where he was heard to be sobbing. VON LAUE and HARTECK went up to see him and tried to comfort him. He appeared to consider himself in the position of a defeated General, the only alternative open to whom is to shoot himself. Fortunately, he had no weapon and he was eventually sufficiently calmed by his colleagues. In the course of conversation with VON LAUE and HARTECK, he made the following remarks:

GERLACH: When I took this thing over, I talked it over with HEISENBERG and HAHN, and I said to my wife: "The war is lost and the result will be that as soon as the enemy enter the country I shall be arrested and taken away". I only did it because, I said to myself, this is a German affair and we must see that German physics are preserved. I never for a moment thought of a bomb but I said to myself: "If HAHN has made this discovery, let us at least be the first to make use of it". When we get back to Germany we will have a dreadful time. We will be looked upon as the ones who have sabotaged everything. We won't remain alive long there. You can be certain that there are many people in Germany who say that it is our fault. Please leave me alone.

6. A little later, HAHN went up to comfort GERLACH when the following conversation ensued:

HAHN: Are you upset because we did not make the uranium bomb? I thank God on my bended knees that we did not make an uranium bomb. Or are you depressed because the Americans could do it better than we could?
GERLACH: Yes.
HAHN: Surely you are not in favour of such an inhuman weapon as the uranium bomb?
GERLACH: No. We never worked on the bomb. I didn't believe that it would go so quickly. But I did think that we should do everything to make the sources of energy and exploit the possibilities for the future. When the first result, that the concentration was very increased with the cube method, appeared I spoke to SPEER's right hand man, as SPEER was not available at the time, an Oberst GEIST(?) first, and later SAUCKEL at WEIMAR asked me: "What do you want to do with these things?" I replied: "In my opinion the politician who is in possession of such an engine can achieve anything he wants". About ten days or a fortnight before the final capitulation, GEIST(?) replied: "Unfortunately we have not got such a politician".

HAHN: I am thankful that we were not the first to drop the uranium bomb.

GERLACH: You cannot prevent its development. I was afraid to think of the bomb, but I did think of it as a thing of the future, and that the man who could threaten the use of the bomb would be able to achieve anything. That is exactly what I told GEIST, SAUCKEL and MURR. HEISENBERG was there at STUTTGART at the time. (Enter HARTECK). Tell me, HARTECK. isn't it a pity that the others have done it?

HAHN: I am delighted.

GERLACH: Yes, but what were we working for?

HAHN: To build an engine, to produce elements, to calculate the weight of atoms, to have a mass-spectrograph and radio-active elements to take the place of radium.

HARTECK: We could not have produced the bomb but we would have produced an engine and I am sorry about that. If you had come a year sooner.

GERLACH: We might have done it, if not with heavy water, then with low temperatures. But when you came it was already too late. The enemy's air superiority was too great and we could do nothing.

HAHN, GERLACH and HARTECK go on to discuss their position if they return to Germany and GERLACH considers that they will have to remain here another two years because they will be in danger. HAHN however feels that he could return to Germany without any danger to himself. GERLACH goes on to explain that the Nazi party seemed to think that they were working on a bomb and relates how the Party people in MUNICH were going round from house to house on the 27th or 28th April last telling everyone that the atomic bomb would be used the following day.

…

7. HAHN and HEISENBERG discussed the matter alone together. HAHN explained to HEISENBERG that he was himself very upset about the whole thing. He said he could not really understand why GERLACH had taken it so badly. HEISENBERG said he could understand it because GERLACH was the only one of them who had really wanted a German victory, because although he realised the crimes of the Nazis and disapproved of them, he could not get away from the fact that he was working for GERMANY. HAHN replied that he too loved his country and that strange as it might appear, it was for this reason that he had hoped for her defeat. HEISENBERG went on to say that he thought the possession of the uranium bomb would strengthen the position of the Americans vis-a-vis the Russians. They continued to discuss the same theme as before that they had never wanted to work on a bomb and had been pleased when it was decided to concentrate everything on the engine. HEISENBERG stated that the people in Germany might say that they should have forced the authorities to put the necessary means at their disposal and to release 100,000 men in order to make the bomb and he feels himself that had they been in the same moral

position as the Americans and had said to themselves that nothing mattered except that HITLER should win the war, they might have succeeded, whereas in fact they did not want him to win. HAHN admitted however that he had never thought that a German defeat would produce such terrible tragedy for his country. They then went on to discuss the feelings of the British and American scientists who had perfected the bomb and HEISENBERG said he felt it was a different matter in their case as they considered HITLER a criminal. They both hoped that the new discovery would in the long run be a benefit to mankind. HEISENBERG went to speculate on the uses to which AMERICA would put the new discovery and wondered whether they would use it to obtain control of RUSSIA or wait until STALIN had copied it. They went on to wonder how many bombs existed. The following is the text of this part of the conversation:

HAHN: They can't make a bomb like that once a week.
HEISENBERG: No. I rather think HARTECK was right and that they have just put up a hundred thousand mass-spectrographs or something like that. If each mass-spectrograph can make one milligramme a day, then they have got a hundred grammes a day.
HAHN: In 1939 they had only made a fraction of a milligramme. They had then identified the "235" through its radio-activity.
HEISENBERG: That would give them 30 k a year.
HAHN: Do you think they would need as much as that?
HEISENBERG: I think so certainly. but quite honestly I have never worked it out as I never believed one could get pure "235". I always knew it could be done with "235" with fast neutrons. That's why "235" only can be used as an explosive. One can never make an explosive with slow neutrons, not even with the heavy water machine, as then the neutrons only go with thermal speed, with the result that the reaction is so slow that the thing explodes sooner, before the reaction is complete. It vaporises at 5000° and then the reaction is already—
HAHN: How does the bomb explode?
HEISENBERG: In the case of the bomb it can only be done with the very fast neutrons. The fast neutrons in 235 immediately produce other neutrons so that the very fast neutrons which have a speed of— say—1/30th that of light make the whole reaction. Then of course the reaction takes place much quicker so that in practice one can release these great energies. In ordinary uranium a fast neutron nearly always hits 238 and then gives no fission.
HAHN: I see, whereas the fast ones in the 235 do the same as the 238, but 130 times more.

HEISENBERG: Yes. If I get below 600,000 V I can't do any more fission on the 238, but I can always split the 235 no matter what happens. If I have pure 235 each neutron will immediately beget two children and then there must be a chain reaction which goes very quickly. Then you can reckon as follows. One neutron always makes two others in pure 235. That is to say that in order to make 10^{24} neutrons I need 80 reactions one after the other. Therefore, I need 80 collisions and the mean free path is about 6 cm. In order to make 80 collisions, I must have a lump of a radius of about 54 cm and that would be about a ton.

HAHN: Wouldn't that ton be stronger than 20,000 tons of explosive?

HEISENBERG: It would be about the same. It is conceivable that they could do it with less in the following manner. They would take only a quarter of the quantity but cover it with a reflector which would turn back the fast neutrons. For instance, lead or carbon and in that way they could get the neutrons which go out, to come back again. It could be done in that way. It is possible for them to do it like that.

HAHN: How can they take it in an aircraft and make sure that it explodes at the right moment'?

HEISENBERG: One way would be to make the bomb in two halves, each one of which would be too small to produce the explosion because of the mean free path. The two halves would be joined together at the moment of dropping when the reaction would start. They have probably done something like that.

HEISENBERG went on to complain bitterly that GOUDSMIT had lied to them very cleverly and thinks that he must at least have told him that their experiments in AMERICA were further advanced. They agreed that the secret was kept very well. HAHN remarked on the fact that there had been no publication of work on uranium fission in British or American scientific journals since January, 1940, but he thought that there had been one published in RUSSIA on the spontaneous fission of uranium with deuterons. HEISENBERG repeated all his arguments saying that they had concentrated on the uranium engine, had never tried to make a bomb and had done nothing on the separation of isotopes because they had not been able to get the necessary means for this. He repeated his story of the alleged threat by America to drop a uranium bomb on DRESDEN and said that he had been questioned by Geheimrat SEGNER (?) of the Foreign Office about this possibility. The conversation concluded as follows:

HEISENBERG: Perhaps they have done nothing more than produce 235 and make a bomb with it. Then there must be any number of scientific matters which it would be interesting to work on.

HAHN: Yes. But they must prevent the Russians from doing it.

HEISENBERG: I would like to know what STALIN is thinking this evening. Of course they have got good men like LANDAU, and these people can do it too. There is not much to it if you know the fission. The whole thing is the method of separating isotopes.

HAHN: No. in that respect the Americans and in fact all the Anglo-Saxons are vastly superior to them. I have a feeling that the Japanese war will end in the next few days and then we will probably be sent home fairly soon and everything will be much easier than it was before. Who knows that it may not be a blessing after all?

8. The guests decided among themselves that they must not outwardly show their concern. In consequence they insisted on playing cards as usual till after midnight. VON WEIZSACKER. WIRTZ, HARTECK, and BAGGE remained behind after the others had gone to bed. The following conversation took place:

BAGGE: We must take off our hat to these people for having the courage to risk so many millions.

HARTECK: We must have succeeded if the highest authorities had said "We are prepared to sacrifice everything".

WEIZSACKER: In our case even the scientists said it couldn't be done.

BAGGE: That's not true. You were there yourself at that conference in Berlin. I think it was on 8 September that everyone was asked— GEIGER. BOTHE and you HARTECK were there too—and everyone said that it must be done at once. Someone said "Of course it is an open question whether one ought to do a thing like that". Thereupon BOTHE got up and said "Gentlemen. it must be done." Then GEIGER got up and said "If there is the slightest chance that it is possible—it must be done." That was on 8 September '39.

WEIZSACKER: I don't know how you can say that. 50% of the people were against it.

HARTECK: All the scientists who understood nothing about it, all spoke against it, and of those who did understand it, one third spoke against it. As 90% of them didn't understand it 90% spoke against it. We knew that it could be done in principal, but on the other hand we realised that it was a frightfully dangerous thing.

BAGGE: If the Germans had spent 10 milliard [billion] marks on it and it had not succeeded, all physicists would have had their heads cut off.

WIRTZ: The point is that in Germany very few people believed in it. And even those who were convinced it could be done did not all work on it.

HARTECK:	For instance, when we started that heavy water business, the CLUSIUS method was apparently too expensive, but I told ESAU that we should use various methods all at once: there was the one in NORWAY, and that we should have a CLUSIUS plant to produce 2–300 l a year; that is a small one and then a hot-cold one. As far as I can see we could never have made a bomb, but we could certainly have got the engine to go.
WIRTZ:	KORSCHING is really right when he said there wasn't very good co-operation in the uranium group as GERLACH said. GERLACH actually worked against us. He and DIEBNER worked against us the whole time. In the end they even tried to take the engine away from us. If a German Court were to investigate the whole question of why it did not succeed in Germany it would be a very, very dangerous business. If we had started properly in 1939 and gone all out everything would have been alright.
HARTECK:	Then we would have been killed by the British "Secret Service".
WIRTZ:	I am glad that it wasn't like that otherwise we would all be dead.

...

At this point HEISENBERG joined WIRTZ and VON WEIZSACKER. The following remarks were passed:

WIRTZ:	These fellows have succeeded in separating isotopes. What is there left for us to do?
HEISENBERG:	I feel convinced that something will happen to us in the next few days or weeks. I should imagine that we no longer appear to them as dangerous enemies.
WEIZSACKER:	No, but the moment we are no longer dangerous we are also no longer interesting. It appears that they can get along perfectly well by themselves.
HEISENBERG:	Perhaps they can learn something about heavy water from us. But it can't be much—they know everything.
WEIZSACKER:	Our strength is now the fact that we are "un-Nazi".

...

III. 7 August

1. On the morning of 7 August the guests read the newspapers with great avidity. Most of the morning was taken up reading these.
2. In a conversation with DIEBNER. BAGGE said he was convinced it had been done with mass-spectrographs.
3. HAHN, HEISENBERG and HARTECK discussed the matter in the following conversation:

HAHN: What can one imagine happens when an atomic bomb explodes'?
                    Is the fission of uranium 0.1, 1, 10 or 100%?
HEISENBERG: If it is 235, then for all practical purposes it is the whole lot, as
                    then the reaction goes much quicker than the vaporisation as for
                    all practical purposes it goes with the speed of light. In order to
                    produce fission in 10^{25} atoms you need 80 steps in the chain so
                    that the whole reaction is complete in 10^{-8} s. Then each neutron
                    that flies out of one atom makes two more neutrons when it hits
                    another uranium 235. Now I need 10^{25} neutrons and that is 2^{80}.
                    I need 80 steps in the chain and then I have made 2^{80} neutrons.
                    One step in the chain takes the same time as one neutron to go
                    5 cm, that is, 10^{-8} s, so that I need about 10^{-8} s. so that the
                    whole reaction is complete in 10^{-8} s. The whole thing probably
                    explodes in that time.
                    (Pause)
HEISENBERG: They seem to have made the first test only on 16 July.
HAHN: But they must have had more material then. They could not make
                    a 100 kg of new uranium 235 every fortnight.
HEISENBERG: They seem to have had two bombs, one for the test and the other
                    for—
HARTECK: But in any case the next one will be ready in a few months.
                    STALIN's hopes of victory will have been somewhat dashed.
HAHN: That's what pleases one about the whole thing. If Niels BOHR
                    helped, then I must say he has gone down in my estimation.

...

5. In a conversation with VON LAUE, VON WEIZSACKER said it will not be
    long before the names of the German scientists appear in the newspapers and
    that it would be a long time before they would be able to clear themselves in the
    eyes of their own countrymen. He went on to quote from the newspaper that we
    were unable to control the energy, from which he assumed that we were not yet
    in possession of a uranium engine, so that their work would still be of con-
    siderable value. He ended by saying:

WEIZSACKER: History will record that the Americans and the English made a
                    bomb. and that at the same time the Germans, under the
                    HITLER regime, produced a workable engine. In other words,
                    the peaceful development of the uranium engine was made in
                    GERMANY under the HITLER regime, whereas the Americans
                    and the English developed this ghastly weapon of war.

...

8. At 6 o'clock the guests all heard Sir John Anderson speak on the wireless. The subsequent conversation was merely a repetition of previous ones. and was chiefly concerned with somewhat caustic comments on the usage to which the new discovery had been put. HEISENBERG's final comment was:

HEISENBERG: If the Americans had not got so far with the engine as we did— that's what it looks like—then we are in luck. There is a possibility of making money.

...

10. HEISENBERG, VON WEIZSACKER, WIRTZ and HARTECK also discussed the future and came to the conclusion that they would probably be sent back to HECHINGEN and allowed to continue their work. They realised however that we might be afraid of their telling the Russians too much. In this connection they mentioned that BOPP, JENSEN and FLUEGGE could also tell them a lot if they wanted to. They came to the conclusion that GROTH was probably in ENGLAND.

IV. The Memorandum Signed by the Guests

All the guests have been extremely worried about the press reports of the alleged work carried out in GERMANY on the atomic bomb. As they were so insistent that no such work had been carried out, I suggested to them that they should prepare a memorandum setting out details of the work on which they were engaged, and that they should sign it. There was considerable discussion on the wording of this memorandum, in the course of which DIEBNER remarked that he had destroyed all his papers, but that there was great danger in the fact that SCHUMANN had made notes on everything. GERLACH wondered whether VOEGLER had also made notes. From the conversation it did however appear that they had really not worked on a bomb themselves, but they did state that the German Post Office had also worked on uranium, and had built a cyclotron at MIERSDORF. ...

Eventually. a memorandum was drawn up and a photostat copy of it is attached to this report. WIRTZ, VON WEIZSACKER, DIEBNER, BAGGE and KORSCHING at first did not want to sign it, but were eventually persuaded to do so by HEISENBERG.

...

Appendix 1

TRANSLATION [of the Memorandum, by the British agents]

8 August 1945

As the press reports during the last few days contain partly incorrect statements regarding the alleged work carried out in Germany on the atomic bomb. we would like to set out briefly the development of the work on the uranium problem.

1. The fission of the atomic nucleus in uranium was discovered by Hahn and Strassmann in the Kaiser Wilhelm Institute for Chemistry in Berlin in December 1938. It was the result of pure scientific research which had nothing to do with practical uses. It was only after publication that it was discovered almost simultaneously in various countries that it made possible a chain reaction of the atomic nuclei and therefore for the first time a technical exploitation of nuclear energies.

2. At the beginning of the war a group of research workers was formed with instructions to investigate the practical application of these energies. Towards the end of 1941 the preliminary scientific work had shown that it would be possible to use the nuclear energies for the production of heat and thereby to drive machinery. On the other hand, it did not appear feasible at the time to produce a bomb with the technical possibilities available in Germany. Therefore, the subsequent work was concentrated on the problem of the engine for which, apart from uranium, heavy water is necessary.

3. For this purpose the plant of the Norsk Hydro at Rjukan was enlarged for the production of larger quantities of heavy water. The attacks on this plan, first by the Commando raid, and later by aircraft, stopped this production towards the end of 1942.

4. At the same time, at Freiburg and later at Celle, experiments were made to try and obviate the use of heavy water by the concentration of the rare isotope U 235.

5. With the existing supplies of heavy water the experiments for the production of energy were continued first in Berlin and later at Haigerloch (Wurtemburg). Towards the end of the war this work had progressed so far that the building of a power producing apparatus would presumably only have taken a short time.

REMARKS [referring to previous paragraphs]

Para. 1. The Hahn discovery was checked in many laboratories, Particularly in the United States, shortly after publication. Various research workers, Meitner and Frisch were probably the first, pointed out the enormous energies which were released by the fission of uranium. On the other hand, Meitner had left Berlin six months before the discovery and was not concerned herself in the discovery.

Para. 2. The pure chemical researches of the Kaiser Wilhelm Institute for Chemistry on the elements produced by uranium fission continued without hindrance throughout the war and were published. The preliminary scientific work on the production of energy mentioned in paragraph 2 was on the following lines:

Theoretical calculations concerning the reactions in mixtures of uranium and heavy water. Measuring the capacity of heavy water to absorb neutrons. Investigation of the neutrons set free by the fission. Investigation of the increase of neutrons in small quantities of uranium and heavy water. With regard to the atomic bomb the undersigned did not know of any other serious research work on uranium being carried out in Germany.

Para. 3. The heavy water production at Rjukan was brought up to 220 l per month, first by enlarging the existing plant and then by the addition of catalytic exchange-furnaces which had been developed in Germany. The nitrogen production of the works was only slightly reduced by this. No work on uranium or radium was done at Rjukan.

Para. 4. Various methods were used for separating isotopes. The Clusius separating tubes proved unsuitable. The ultra-centrifuge gave a slight concentration of isotope 235. The other methods had produced no certain positive result up to the end of the war. No separation of isotopes on a large scale was attempted.

Para. 5. Further a power producing apparatus was prepared which was to produce radio-active substances in large quantities artificially without the use of heavy water but at very low temperatures.

Paras. 3 and 5. On the whole the funds made available by the German authorities (at first the Ordnance Department and later the Reichs Research Board) for uranium were extremely small compared to those employed by the Allies. The number of people engaged in the development (scientists and others. at institutes and in industry) hardly ever exceeded a few hundred.

Signed: Otto Hahn, Walther Gerlach, P. Harteck, K. Wirtz, H. Korsching, M. v. Laue, W. Heisenberg, C. F. v.Weizsäcker, E. Bagge, K. Diebner

(My signature signifies that I share responsibility for the accuracy of the above statement, but not that I took any part whatever in the above mentioned work. Signed- M. v. Laue)

…

From FARM HALL REPORT 5

OPERATION "EPSILON"

(8–22 August, 1945)

1. GENERAL

1. The guests have recovered from the initial shock they received when the news of the atomic bomb was announced. They are still speculating on the method used to make the bomb and their conversations on this subject, including a lecture by HEISENBERG, appear later in this report.

…

III. TECHNICAL

6. The following day [9 August 45] the newspapers mentioned that the atomic bomb weighed 200 k and the following conversation took place between HARTECK and HEISENBERG.

HARTECK: Do you believe that it is true that this is the weight of the bomb or that they wish to bluff the Russians?

HEISENBERG: This has worried me considerably, and therefore this evening I have done a few calculations and have seen that it is more probable than we had thought on account of the substantial multiplication factors which one can have with fast neutrons. We have always calculated with a multiplication factor of 1.1 because we had found this in practice with uranium. If they have a multiplication factor of 3 or 5 then naturally it is a different matter. We said we need about 80 links in the chain reaction; now the mean free path is 4 cms. Therefore we must have 80 long divisions (so was the rough estimate) and this would then come to about a ton. This calculation is right if the multiplication factor is 1.1 because even then we use really every neutron which "escapes" for multiplication. If on the other hand the multiplication factor is 3, things are quite different. Then I can say, if the whole thing is only as big as the mean free path, then one neutron which walks around therein once meets another and makes three neutrons. From these three neutrons one will already come back, the other two can go off; the one that comes back will for certain make another three. In practice therefore I need only the mean free path for the thing to work.

HARTECK: What are these 100–200 k which are around it?

HEISENBERG: This will be the reflector. For instance they might have lead as a reflector and of course part of the weight could be apparatus; you must remember that it is a dangerous business. They must arrange it so that it is at first taken apart into two pieces between which a reflector, of shall we say lead or some other material, is placed, and then at the right moment this lead will be pulled out and the thing clapped together.

                  (Pause). Well how have they actually done it? I find it is a disgrace if we, the Professors who have worked on it, cannot at least work out how they did it.

…

HEISENBERG: I do not believe that the Americans could have done it [obtained element 94, plutonium]. They would have had to have had, shall we say, a machine running at least not later than 1942 and they would have had to have had this machine running for at least a year and then they would have had to have done all this chemistry.

HARTECK: Highly improbable.

…

From FARM HALL REPORT 6

OPERATION "'EPSILON"

(23 August–6 September. 1945)

IV. The Future.

…

2. HEISENBERG made it quite clear to me [Rittner] that he wishes to continue work on uranium, although he realises that this could only be done under Allied control. His main interest at the moment is to get back to GERMANY to look after his family, who appear to be in some difficulty as they live in the mountains near Munich, and his wife has no one to help her with her seven children. He is very distressed to hear from his wife that his mother died two months ago, and that a woman friend of his wife who had been helping her had also died. He is perfectly prepared to give an undertaking on oath not to work on uranium, except under Allied control, if he is allowed to return to his family. HEISENBERG has threatened to withdraw his parole, which he gave me in writing some time ago, unless some arrangement is made regarding the future of himself and his family.

…

From FARM HALL REPORT 18

OPERATION "EPSILON"

(12–18 November 1945)

V. NOBEL PRIZE

9. The "Daily Telegraph's" announcement of the award to HAHN of the Nobel Prize for Chemistry caused general pleasure and also deep misgivings as no official confirmation was forthcoming. It was even thought that some unaccountable malice was responsible for our withholding the news. However, great efforts were made in LONDON to try to verify the report and, as the source seemed reasonably reliable, the award was duly celebrated with songs, speeches, baked meats and some alcohol. Proceedings started very badly with an unfortunate speech by von LAUE, at the end of which, both he and HAHN were in tears to everybody's great discomfort, particularly mine, as I was sitting between them. However, the united efforts of the rest of the party restored our normal good spirits.

10. I hope to obtain copies of the better songs and speeches and to publish them later in the form of an appendix to this report. They will, of course, be of no operational interest.

11. VON LAUE told HAHN that there was a good reason for the award of the prize.

12. LAUE: You have not got the NOBEL prize as a consolation for we ten German scientists who are shut up here, but I think that is their reason for giving it to you now instead of waiting till next year which was probably their original intention.

13. HAHN said that, if he were allowed to go to SWEDEN to receive the prize, he would not be able to give his word to say nothing about his detention here.

14. HARTECK: They (Allies) cannot make you give your word not to say where you have been and with whom.

HAHN: Out of the question!

HARTECK: They won't make you give your word, but they will say: "'Of course, we are delighted to send you to SWEDEN, but please don't say with whom you are".

Excerpt from Erich Bagge's "Internment Diary" (Translated)

Erich Bagge et al. 1957, 66–67. Another translation may be found in Bernstein 2001, 323.

Sunday, 18 November 1945

Today I can report on an interesting event that is doubtless without comparison. On Friday morning, just after breakfast, most of us were sitting in the salon in order to listen to "this week's composer"—it was Rimsky-Korsakov—and to study "the latest news," when Heisenberg said to Hahn, "Mr. Hahn, read this!" And he handed him the *Daily Telegraph*.

Mr. Hahn, who himself was just then intently reading another paper, replied, "I don't have any time right now."

"But this is very important for you. It says here that you are to receive the Nobel Prize for 1944."

The excitement that overcame the 10 detainees is naturally very difficult to describe in a few words. Hahn did not believe it at first. Initially he refused all statements of congratulations. But then we gradually broke through, led by Heisenberg, who heartily congratulated him for the 6200 [British] pounds. Then we followed. Heisenberg immediately went to the Captain, who was completely surprised by the news and still could not comprehend it a half hour later. He immediately called the London Office. Apparently no one there knew anything, but the telephones were soon put into very lively use. Information Ministry, *Times* correspondent and all possible newspaper people were called. It soon became apparent that this was not a false report.

Bibliography

Archival Materials

Alsos Mission Documents. US National Archives and Records Administration (NARA), College Park, MD. Record Group 77.11.3 (Office of Chief of Engineers, Manhattan Engineer District Headquarters), also Microfilm Group M1109.

Farm Hall Reports (U.S. copy). NARA, RG 77.11.1 (Office of the Commanding General), Entry 22, Box 163.

Farm Hall Reports (British copy). British National Archives, Kew, London, reference number WO 208/5019. In PDF: http://discovery.nationalarchives.gov.uk/SearchUI/Details?uri=C4414534

Goudsmit, Samuel A. Papers: AIP NBLA, online at: http://wwwproxy.aip.org/history/nbl/collections/goudsmit/

Laue, Max von. Letters to his son while at Farm Hall. Max von Laue Papers, Deutsches Museum, Munich, 1976-20.

Publications

Aristotle. 1984. Poetics. In *The Complete Works of Aristotle: The Revised Oxford Translation*, vol. 2, ed. Jonathan Barnes, 2316–2340. Princeton: Princeton University Press.

Ball, Philip. 2014. *Serving the Reich: The Struggle for the Soul of Physics under Hitler*. Chicago: University of Chicago Press.

Bagge, Erich, Kurt Diebner, and Kenneth Jay. 1957. *Von der Uranspaltung bis Calder Hall*. Hamburg: Rowohlt.

Bentley, Eric. 1952. Introduction: The Science Fiction of Bertolt Brecht. In Bertolt Brecht. *Galileo*, trans. Charles Laughton, ed. Eric Bentley, 9–42. New York: Grove Press.

Bernstein, Jeremy. (ed.) 2001. *Hitler's Uranium Club: The Secret Recordings at Farm Hall*. New York: Springer (earlier edition. Woodbury, NY: American Institute of Physics Press, 1996).

Beyerchen, Alan. 1977. *Scientists under Hitler: Politics and the Physics Community in the Third Reich*. New Haven: Yale University Press.

Cassidy, David C. 1994. Controlling German Science, I: U.S. and Allied Forces in Germany, 1945–1947. *HSPS: Historical Studies in the Physical and Biological Sciences* 24: 197–235.

Cassidy, David C. 2001. *Introduction*. Bernstein 2001, xvii–xxx; Bernstein 1996, xiii–xxvi.

Cassidy, David C. 2009. *Beyond Uncertainty: Heisenberg, Quantum Physics, and the Bomb*. New York: Bellevue Literary Press (earlier edition: *Uncertainty: The Life and Science of Werner Heisenberg*. New York: W. H. Freeman, 1992).

Dahl, Per F. 1999. *Heavy Water and the Wartime Race for Nuclear Energy*. Bristol (U.K.): Institute of Physics Publishing.

© Springer International Publishing AG 2017

D.C. Cassidy, *Farm Hall and the German Atomic Project of World War II*,
DOI 10.1007/978-3-319-59578-8

Dörries, Matthias. ed. 2005. *Michael Frayn's "Copenhagen" in Debate*. Berkeley: Office for History of Science and Technology, 2005. Contains Bohr's draft letters to Heisenberg in facsimile, transcription, and English translation, 101–179.

Egleton, Clive. 1997. *The Alsos Mission*. London: Severn House Publishers.

Farm, Hall. 1993. *Operation Epsilon: The Farm Hall Transcripts*. No author. Introduction by Sir Charles Frank. Bristol (UK): Institute of Physics Publishing; and Berkeley: University of California Press.

Frayn, Michael. 1998. *Copenhagen*. New York: Anchor Books.

Fry, Helen. 2012. *The M Room: Secret Listeners Who Bugged the Nazis*. London: Marranos Press.

Gimbel, John. 1986. U. S. Policy and the German Scientists: The Early Cold War. *Political Science Quarterly* 101: 433–451.

Goudsmit, Samuel A. 1996. *Alsos*. Woodbury, NY: American Institute of Physics Press; reprint of earlier edition, New York: Henry Schuman, 1947.

Groves, Leslie R. 1983. *Now It Can Be Told: The Story of the Manhattan Project*. New York: DaCapo Press (reprint of earlier edition, New York: Plenum Publishers, 1962).

Heilbron, J. L. 1986. *The Dilemmas of an Upright Man: Max Planck as Spokesman for German Science*. Berkeley: University of California Press.

Heisenberg, Elisabeth. 1984. *Inner Exile: Recollections of a Life with Werner Heisenberg*. S. Cappellari and C. Morris, trans. Boston: Birkhäuser.

Heisenberg, Werner. 1936. Zum Artikel: Deutsche und jüdische Physik. Entgegnung von Prof. Dr. Heisenberg. *Völkischer Beobachter, 49*, no. 59 (28 February 1936), 6. Reprint: Heisenberg 1989b, 10–11. English translation: Hentschel 1996, 121–123.

Heisenberg, Werner. 1942. Die theoretischen Grundlagen für die Energiegewinnung aus der Uranspaltung. Lecture delivered 26 February 1942. Published Heisenberg 1989a, 517–521.

Heisenberg, Werner. 1971. *Physics and Beyond: Encounters and Conversations*. Arnold J. Pomerans, trans. New York: Harper and Row.

Heisenberg, Werner. 1989a. *Gesammelte Werke/Collected Works*. Vol. A2. ed. W. Blum et al. Berlin: Springer.

Heisenberg, Werner. 1989b. *Gesammelte Werke/Collected Works*. Vol. C5. ed. W. Blum et al. Munich: R. Piper.

Heisenberg, Werner. 2003. *Liebe Eltern! Briefe aus kritischer Zeit 1918 bis 1945*, ed. Anna Maria Hirsch-Heisenberg, Munich: Langen Müller.

Heisenberg, Werner and Elisabeth Heisenberg. 2016. *My Dear Li: Correspondence, 1937–1946*. ed. Anna Maria Hirsch-Heisenberg, Trans. Irene Heisenberg, New Haven: Yale University Press.

Hentschel, Klaus, ed. 1996. *Physics and National Socialism: An Anthology of Primary Sources*. Trans. A. M. Hentschel, Basel: Birkhäuser.

Hoffmann, Dieter (trans.). 1993. *Operation Epsilon: Die Farm-Hall-Protokolle oder die Angst der Alliierten vor der deutschen Atombombe*. Berlin: Rowohlt.

Hoffmann, Dieter, and Mark Walker (eds.). 2007. *Physiker zwischen Autonomie und Anpassung: Die Deutsche Physikalische Gesellschaft im Dritten Reich*. Weinheim: Wiley-VCH.

Jones, R.V. 1978. *The Wizard War: British Scientific Intelligence 1939–1945*. New York: Coward, McCann and Geoghegan.

Jungk, Robert. 1958. *Brighter Than a Thousand Suns: A Personal History of the Atomic Scientists*. Trans. James Cleugh, New York: Harcourt Brace Jovanovich (Translation of *Heller als Tausend Sonnen*. Bern: Alfred Scherz Verlag, 1956).

Kramish, Arnold. 1986. *The Griffin*. Boston: Houghton Mifflin.

Mahoney, Leo James. 1981. A History of the War Department Scientific Intelligence Mission (Alsos), 1943–1945. PhD dissertation, Kent State University.

McPartland, Mary A. 2013. The Farm Hall Scientists: The United States, Britain, and Germany in the New Atomic Age, 1945–46. PhD dissertation, The George Washington University.

Pash, Boris T. 1969. *The Alsos Mission*. New York: Award House.

Powers, Thomas. 2000. *Heisenberg's War: The Secret History of the German Atomic Bomb*. New York: DaCapo Press.

Rhodes, Richard. 1986. *The Making of the Atomic Bomb*. New York: Simon and Schuster.

Rose, Paul Lawrence. 1998. *Heisenberg and the Nazi Atomic Bomb Project*. Berkeley: University of California Press.

Shepherd-Barr, Kirsten. 2006. *Science on Stage: From* Dr. Faustus *to* Copenhagen. Princeton: Princeton University Press.

Sime, Ruth Lewin. 1996. *Lise Meitner: A Life in Physics*. Berkeley: University of California Press.

Walker, Mark. 1989. *German National Socialism and the Quest for Nuclear Power, 1939–1949*. New York: Cambridge University Press.

Walker, Mark. 1995. *Nazi Science; Myth, Truth, and the German Atomic Bomb*. New York: Plenum Publishers.

Printed in the United States
By Bookmasters